THE SERVICE CONSULTANT

Working in an Automotive Facility

THE SERVICE CONSULTANT

Working in an Automotive Facility

Ron Garner, Ph.D.

Pennsylvania College of Technology

Williamsport, Pennsylvania

C. William Garner, D.Ed.

Rutgers University

New Brunswick New Jersey

THOMSON

DELMAR LEARNING

Australia • Canada • Mexico • Singapore • Spain • United Kingdom • United States

THOMSON
DELMAR LEARNING

The Service Consultant: Working in an Automotive Facility

Ronald Garner, Ph.D., and C. William Garner, D.Ed.

Vice President, Technology and Trades SBU:

Alar Elken

Editorial Director:

Sandy Clark

Senior Acquisitions Editor:

David Boelio

Developmental Editor:

Matthew Thouin

Marketing Director:

David Garza

Channel Manager:

William Lawrensen

Marketing Coordinator:

Mark Pierro

Production Director:

Mary Ellen Black

Production Editor:

Barbara L. Diaz

Technology Project Manager:

Kevin Smith

Technology Project Specialist:

Linda Verde

Editorial Assistant:

Andrea Domkowski

Library of Congress
Cataloging-in-Publication Data:

Garner, Ron, 1966–
    The service consultant : working in an automotive facility / Ron Garner, C. William Garner.
    p. cm.
    ISBN 1-4018-7990-X (alk. paper)
    1. Service stations—Management.
2. Automobiles—Maintenance and repair. 3. Consultants. 4. Automobile industry and trade—Customer services.
I. Garner, C. William.   II. Title.

TL153.G365 2005
629.28′6′068—dc22   2004058820

NOTICE TO THE READER

Publisher does not warrant or guarantee any of the products described herein or perform any independent analysis in connection with any of the product information contained herein. Publisher does not assume, and expressly disclaims, any obligation to obtain and include information other than that provided to it by the manufacturer.

The reader is expressly warned to consider and adopt all safety precautions that might be indicated by the activities herein and to avoid all potential hazards. By following the instructions contained herein, the reader willingly assumes all risks in connection with such instructions.

The publisher makes no representation or warranties of any kind, including but not limited to, the warranties of fitness for particular purpose or merchantability, nor are any such representations implied with respect to the material set forth herein, and the publisher takes no responsibility with respect to such material. The publisher shall not be liable for any special, consequential, or exemplary damages resulting, in whole or part, from the readers' use of, or reliance upon, this material.

CONTENTS

PREFACE

This book is for people who wish to work in an automotive facility as a service consultant. The authors recognize that the position of service consultant must be examined from different perspectives and in reference to a variety of relationships. The title "service consultant" is used in this book to recognize the numerous tasks that are much broader in scope than those assigned to a person occupying a position of customer service.

First of all, the authors recognize that the tasks and procedures assigned to a service consultant will vary dependent on whether he or she works for an independent automotive service facility, a dealer representing an automotive manufacturer, a franchise, or a private or public fleet. Secondly, the authors have used the word "service" to represent a facility that either provides for the maintenance of automobiles and/or for their repair. Therefore, a service consultant may work in a facility that offers both or either of the services to automobile owners.

Second, the book attempts to cover the numerous tasks that may be performed by a service consultant. Admittedly, some service consultants are required to perform a wide range of tasks while others are given a limited number of tasks. The effort to present as many tasks as possible is intended to prepare the student for a variety of service consultant positions plus provide the ability to move from one type of service facility to another or from an entry-level position to one that affords greater responsibility.

Thirdly, the book addresses the need for the service consultant to use computers to manage customer files. From personal experience, the authors know the importance of using a computer to run a service facility. Because computer software programs are quite different from each other, the actual computer entries and commands must be learned from the software manufacturer. This book, however, thoroughly covers the application of the computer in the tasks to be performed by the service consultant.

Finally, throughout the book, the importance of the customer is stressed. In some cases, automobile owners come to a facility with a clear-cut knowledge of what has to be done to their automobile. In other cases, the owners depend on the service consultant to assist them. From the authors' experiences, dependence on the service consultant for advice is the growing trend. Thus, the book's approach to customer relations is that the service consultant should offer customers professional assistance.

Book Overview

Chapters 1 through 3 focus on the different types of service facilities, the tasks and duties of the service consultant, and the personnel involved in the processing of work in a service facility. Chapters 4 through 8 move to the customer process from the initial contact to the presentation of the invoice. Chapters 9 through 12 fill in the details of the actions and procedures described in previous chapters. Chapters 13 and 14 focus on how the general operations and the work environment in an automotive facility can affect a service consultant's job and the attitudes of other employees.

The chapter content is aligned with tasks identified by Automotive Service Excellence (ASE) for the Automotive Service Consultant. Each chapter begins with a list of learning objectives, many of which directly reference an ASE task, and ends with a set of exercises related to the learning objectives. In addition, at the end of each section, suggested activities offer students some insights into the operations of different service facilities.

From the Authors

The content in this book is largely based on our experiences as owners/operators of an automotive service facility; our studies of automotive technology, business, and vocational education; and our work as teachers/trainers and consultants. Starting an automotive service facility in an empty building with two old lifts and a compressor, we gained invaluable insights into the trials and tribulations associated with creating a business from the ground up.

As owners/operators, we did everything, working as service consultant, cashier, manager, technician, and even custodian.

Most of what we talk about in this book we have done. We try to present as honest a picture as we can and hope it will benefit those of you who will enter the automobile industry. If you like being around cars and people, you will enjoy the job of a service consultant.

ABOUT THE AUTHORS

Ron Garner, Ph.D. is an Associate Professor of Automotive Technology in the Division of Transportation Technology at the Pennsylvania College of Technology, an affiliate of Pennsylvania State University. In that capacity he oversees a baccalaureate program in Automotive Technology Management, teaches a wide range of upper level technology courses, teaches Enhanced Pennsylvania Emission Certification, directs baccalaureate student final projects, and consults with organizations in business and nonprofit sectors.

He completed an A.A.S. degree in Ford ASSET automotive technology from Lehigh County Community College, a B.S. in Vocational Education, and an M.S. in Vocational Education with an emphasis in School Administration and Leadership from Pennsylvania State University. He earned his doctorate in Workforce Education at Pennsylvania State University with an emphasis in Industrial Training and Post Secondary School Administration. An A.S.E. certified Master Automobile Technician with the Advanced Engine Performance and Compressed Natural Gas Certification, he has been a technician as well as owned and managed automobile repair facilities.

C. William Garner, D.Ed. served as an airborne sonar technician in the U.S. Navy from 1959 to 1963. He earned a bachelor's degree in business education, a master's degree in higher education administration, and a doctorate in vocational education at Pennsylvania State University. He then took an appointment with Southern Illinois University at Carbondale as an Assistant Professor of Occupational Education and Site Administrator at March Air Force Base in California. His next appointment was with the University of Louisville as an Assistant Professor of Vocational Education and Coordinator for Educational Programs at Fort Knox.

In 1978 he received an appointment as an Associate Professor of Vocational Education at Rutgers University. As a professor at Rutgers he has served as Chair of the Department of Urban Education, Chair of the Graduate Department of Vocational Education, Executive Director of the Vocational Education Resource Center, and Acting Dean. Currently, Dr. Garner is an Associate Professor of Education Administration at Rutgers University and has written books on government accounting and school finance. In addition, he and his son, Ron, owned and operated an auto repair business.

PART I

SERVICE FACILITIES
AND THE
SERVICE CONSULTANT

CHAPTER 1

TYPES OF AUTOMOTIVE SERVICE FACILITIES

OBJECTIVES

Upon reading this chapter, you should be able to:

- *Describe the different types of automobile service facilities.*

- *Diagram an organizational structure for the different types of automobile service facilities.*

- *Explain the importance of the guidelines that service facility operations must follow in different states.*

- *Identify the major financial measures that have a direct impact on the profit or loss of a service facility.*

Introduction

A clear understanding of the terms used in a book is critical. For that reason, this chapter begins by defining two important terms used throughout the book. First, the term **service** means the maintenance, repairs, and diagnosis of an automobile. Second, an **automobile service facility** is a for-profit business that services automobiles.

The purpose of this chapter is to point out several important initial distinctions that service consultants should make among different auto service facilities. First, not all automobile service facilities are the same. While the job of the service consultant is similar at most service facilities, there are important differences. Being aware of these differences is critical for service consultants.

Second, automobile service facilities have different organizational structures. An **organizational structure** represents the managerial chain of command. An organizational diagram presents the relationships among the positions in an automobile service facility. A position that is higher in the diagram is responsible for the supervision and evaluation of the position beneath it. Knowing to whom the service consultant reports is critical.

In addition, service consultants must be knowledgeable about the state laws and directives that regulate the operation of their service facility, the people who work in the facility, and the automobiles serviced. This chapter provides a discussion of the regulations found in many states.

Finally, this book is not a management text and does not expect the service consultant to be an accountant. Therefore, it does not present an in-depth discussion of the financial operations of a service facility. However, the service consultant must have an awareness of the bottom line, meaning the generation of a profit for the facility. Regardless of the effectiveness of the person serving as consultant, a service facility cannot stay in business if it does not make a profit. This chapter presents an introduction to basic financial measures that are important to the service consultant.

Types of Automotive Facilities by System or Product

An automobile service facility sells the labor of a trained automotive technician to one customer at a time. The objective of an automobile service facility is to solve a customer's automotive problems. To achieve this objective the service facility may have to perform a diagnosis to identify the problem, repair or replace a part, and/or perform maintenance to keep the automobile in top condition.

In most automobile service facilities, the service consultant is the person who works directly with the customer and technicians to arrange for the diagnosis, repair, or maintenance of the automobile. As a result, service consultants must have the ability to communicate with other people and have a thorough knowledge of the types of repairs and maintenance offered by their service facility.

Because the automobile is the most complex machine a person owns, some service facilities specialize in certain services. One of the mistakes a service consultant in a specialty service facility can make is to accept work that is not done by his or her service facility technicians.

Automobile Systems

One way to classify the types of automobile service facilities is by the automobile system or systems that the facility's technicians service. Eight different automobile systems are listed by the National Institute for Automotive Service Excellence (ASE):

- Engine Repair (A1)
- Automatic Transmission/Transaxle (A2)
- Manual Drive Train and Axles (A3)
- Suspension and Steering (A4)
- Brakes (A5)
- Electrical/Electronic Systems (A6)
- Heating and Air Conditioning (A7)
- Engine Performance (A8)

Naturally, when a facility services one of these automobile systems, it must employ automobile technicians with the expertise needed to diagnose, repair, and maintain the system on the various makes and models of automobiles owned by customers.

One service facility may work on all or most of the systems listed above. As shown in the picture in Figure 1-1, such a facility must have the

FIGURE 1-1 An automobile service facility that works on multiple systems.

FIGURE 1-2 A system-specific service facility.

space to house the different equipment, inventory, and tools needed to work on multiple systems as well as the equipment and space needed by the service consultants. In the service facility that works on multiple systems, the service consultants must be thoroughly familiar with all of the systems. In addition, they must be knowledgeable about the parts used to make the repairs, the specific expertise of each of the technicians, and the means to maintain accurate records of customers, services, and parts.

Some automobile service facilities repair and maintain one system, such as transmissions or brakes. Such a service facility may be referred to as a **system-specific service facility**. As shown in the picture in Figure 1-2, the facility, equipment, and tools used by a muffler and brake specialty shop are limited to muffler and brake diagnosis, repairs, and maintenance.

Naturally, a facility that limits its work to one or two systems must have technicians and service consultants who are thoroughly familiar with the specific system or systems being serviced. They must understand the system's repair procedures, the parts needed to make the repairs, and the related maintenance procedures.

Product-Specific Service Facilities

Another type of service facility diagnoses, repairs, and maintains a particular make and model of automobile. This service facility is referred to as a product-specific service facility. For example, a facility might specialize in the diagnosis, repair, and maintenance of Volkswagens (see Figure 1-3).

The most common product-specific service facility is an automobile dealership that sells new automobiles. Although it specializes in the models it sells, its service facility may provide similar repairs and maintenance for

FIGURE 1-3 A product-specific service facility.

other makes of automobiles. At a dealership, the service consultant must be thoroughly informed about the different models sold as well as the abilities (and sometimes preferences) of the technicians who perform the work.

Facility Ownership

Automobile service facilities differ in ownership. For example, one person, several people, or a corporation may own a service facility. A **proprietorship** means that one person owns a service facility, while a **partnership** means two or more people are the owners. If an owner or the owners wish, they may incorporate the service facility. A **corporation** may be owned by one or more persons.

A person's ownership in a corporation depends on the amount of money the person invests. People invest in a corporation by purchasing shares of stock and so are called the stockholders. Their percentage of corporate ownership depends on the number of stock shares they purchase. Because a corporation becomes a legal entity, service facilities are incorporated for legal and financial benefits.

In all of the types of ownerships (proprietorship, partnership, and corporation), the money invested by the owners or stockholders is used to prepare the service facility to conduct business; for example, the purchase of land, buildings, tools, equipment, inventory, and furniture.

In a new service facility, invested money is usually needed to run the business for a short period until it can earn a profit. The profit earned by the service facility goes to the owners or stockholders, and the amount of the profit an investor receives depends on the amount the person invested.

In addition, the service facility may be classified as a:

- privately owned service facility
- franchise
- chain
- dealership
- fleet service department

Obviously, the method of ownership influences the job activities of the service consultant and the number and type of people employed by the service facility. In addition, the type of ownership may influence the management structure of the facility. For example, who owns the facility and what role does the owner play in its day-to-day operations?

Privately Owned Service Facility

A privately owned facility may provide services to all automobile systems or limit its services to one system or product. The owners may change the systems to be repaired, services offered, or products sold at their own discretion.

Figure 1-4 presents an organizational structure for a private proprietorship owned by Mr. Williams. As shown in this diagram, Mr. Williams is in the top box and serves as the general manager and service manager of the service facility. The service consultant (Kevin), the lead technician (Rich), and the parts specialist (Bob) report directly to the owner-general/service manager.

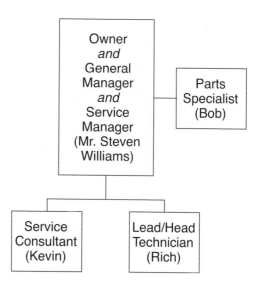

FIGURE 1-4 An organizational structure for a proprietorship or partnership.

In other words, Mr. Williams is responsible for the hiring, direct supervision, and evaluation of the work of the service consultant, the technician, and the parts specialist. The service consultant and lead technician work directly with each other on a daily basis but neither is subordinate to the other. Likewise, the service consultant and lead technician interact directly with the parts specialist on a daily basis. In other words, the diagram represents a chain of command and not a diagram for communications or working relationships.

If more technicians are employed they would be under the supervision of the lead technician, and if people were hired in the parts department they would report to the parts specialist. Each would supervise and assign work to his or her subordinates, but decisions to hire or terminate them would be made by the owner unless the owner delegates this authority to the service consultant, lead technician, or parts specialist.

Corporate Owned Facilities

Assume that Mr. Williams decided to change his business from a proprietorship to a corporation and sell stock in his business to investors. Figure 1-5 presents a diagram for a corporate owned service facility. Note that the name in the top box is "stockholders" as opposed to an owner or owners. Williams now reports to the stockholders and also serves as the general manager.

In Figure 1-5, Mr. Williams is the immediate supervisor of the four managers. As the general manager, Mr. Williams relies on the service man-

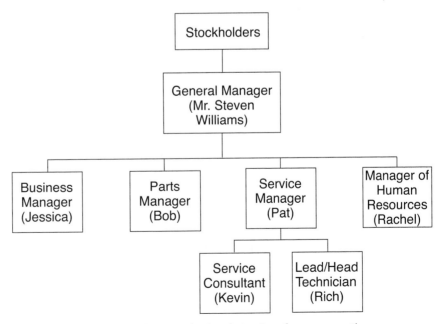

FIGURE 1-5 An organizational structure for a corporation.

ager (Pat), business manager (Jessica), manager of human resources (Rachel), and parts manager (Bob) to assist him with the operations of the business. This assistance includes working with the day-to-day operations of the service facility, overseeing contracts with and deliveries to parts stores, recruiting new employees, working with the local bank on the checking account and loans, ensuring availability of telephone service, approving computer software purchases, and many other activities.

The four managers work as a team under the general manager. Also, Figure 1-5 shows that the service manager supervises the service consultant and the lead technician. Because the diagram in Figure 1-5 exhibits the managerial chain of command and not the interactions between employees, the lead technician or service consultant orders parts through the parts manager, and service orders are exchanged between the service consultant and lead technician. **Vendor invoices** are received by the business manager, who works with the parts manager to verify the delivery of parts. The manager of human resources and business manager work together on the payroll.

In Figure 1-5 the service manager has direct supervision of the service consultant and the lead technician. This implies that Mr. Williams has given the service manager the authority to hire, supervise, and evaluate the performances of the service consultant and lead technician. Further, the service manager works directly with the human resources manager to recruit new employees and with the business manager to get them on the payroll. The service manager reports his actions to the general manager or presents a recommendation to the general manager for approval. Whether or not the service manager takes action and reports or makes a recommendation for approval is up to the general manager. At the same time, if there are personnel in the parts and other departments, the assumption from the diagram in Figure 1-5 is that, while the managers are responsible for the day-to-day operations, the general manager is personally involved in the hiring, supervision, and evaluation of employees.

Franchise Ownerships

Because the startup and operation of a service facility can be expensive and complicated, some people who wish to own a service facility may choose to purchase a franchise. A **franchise** permits an owner to use a nationally recognized name and receive some assistance, such as training, advertising, and consulting, for startup operations.

People wishing to start a franchise usually have to submit a business plan when applying to purchase the rights to open the service facility. They must also have the money to purchase the franchise, buy the necessary equipment and supplies, and obtain the building space to house the operation. The national franchise corporation that owns the name of the franchise (such as Midas Muffler, AAMCO, or Tune and Lube, among others) has to approve the request presented by the people making application as well as all local arrangements, such as the location and building.

In return for the purchase of a franchise, the national franchise corporation typically requires a percentage of the gross sales received by the local service facilities. In addition, it usually has the right to conduct periodic inspections of the franchise's operations and records and may close a local service facility as warranted under the contract.

In most cases, the national franchise corporation requires that the local owner and service consultant attend its training programs and learn to use its forms to record all local services and receipts. The training of service consultants may require that they read written scripts when talking to customers on the phone or communicating with a customer in person.

An organizational diagram of a franchise is shown in Figure 1-6. Note that in this diagram the national franchise corporation is in a supervisory position similar to the stockholders in Figure 1-5. As in Figure 1-5, Mr. Steven Williams is the local owner, general manager, and service manager, with Kevin working as the service consultant and Rich as the lead technician. In a larger franchise, the corporation may prefer that a person other than the owner serve as the service manager and may choose to hire multiple managers such as those shown in Figure 1-5 (a business manager, several technicians, a parts manager or specialist, and a human resources manager).

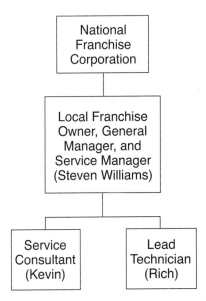

FIGURE 1-6 The organizational structure of a franchise service facility.

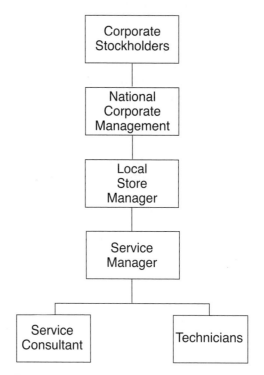

FIGURE 1-7 The organizational structure of a chain owned service facility.

Chain Ownerships

A service facility that is one of several facilities owned by a corporation is called a **chain service facility** (such as Sears, Roebuck & Co, Wal-Mart, Pep Boys, and many others). The chain service facilities sell the same services and use the same procedures regardless of location.

Each service facility has a service manager, at least one service consultant, and several technicians. As shown in Figure 1-7, a service manager usually reports to the store manager, who in turn reports to a manager in a national corporate office. The owners of the corporation's chain of service facilities are the corporate stockholders, who provide the money needed to buy the tools, equipment, and building that houses the local service facility.

Automobile Dealership

An automobile dealership sells new and used automobiles as well as the parts needed to repair and maintain them. The dealership service department is considered a **specialty service facility** because its primary objective is to diagnose, repair, and maintain the automobiles sold by the

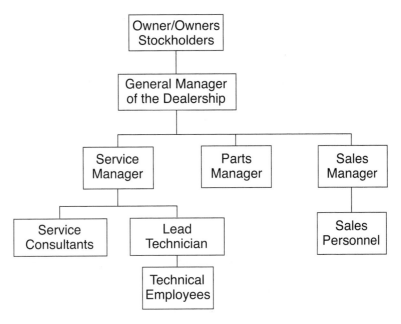

FIGURE 1-8 The organizational structure for an automobile dealership.

dealership. Actually, a dealership is really a franchise of an automobile manufacturer since the manufacturer approves the dealership, can close it if warranted, and makes money from the selling of the automobiles and replacement parts (called a product franchise).

Figure 1-8 shows a diagram with the owners/stockholders at the top. One or more people may own a dealership; however, it is usually incorporated. Under the owners/ stockholders is a general manager, president, or vice president who is responsible for a variety of departments or divisions in the dealership.

In Figure 1-8, three departments are shown, although a dealership typically contains more departments and managers such as a business manager, a manager for the maintenance of the buildings and grounds, and so on. With respect to the service facility, a service manager is shown as the person in charge of the service operations. Reporting to the service manager are the service consultants and the lead technician.

It is important to note again that Figure 1-8 is an example that shows how an organizational structure may appear at an automobile dealership. The service consultants, service managers, and technicians may be in different positions in a dealership diagram, but the point is that the individuals occupying these positions must know to whom they report. The employees must know who is assigned to supervise and evaluate their performance.

Fleet Service Departments

Companies with fleets of automobiles often have a fleet service department. The **fleet service facility** employs automobile technicians who diagnose, repair, and maintain its fleet of automobiles. Unlike the other service facilities, the fleet service department does not service customer vehicles but rather the vehicles owned by the company. Even though there are no customers in the traditional sense, a service manager is usually employed and expected to keep the vehicles in the fleet in top condition and the cost of servicing the fleet within an assigned budget.

State Guidelines for Service Facility Operations

Some states require that automobile service facilities obtain a state issued license to sell services to customers. Service consultants must be thoroughly informed about these legal requirements. For example, California's Bureau of Automotive Repair must issue a license to an automobile service facility before it can sell automobile repairs to customers. In addition, in some states, a technician may have to obtain an automobile technician's license before he or she can provide automobile services to a customer. An example of this regulation is New York's Automotive Technician License.

In other states a service facility and its technicians may need a license before either can perform certain types of work. For example, in Pennsylvania, automobiles must be inspected every year to make sure they meet state safety regulations. Pennsylvania also requires that automobile technicians obtain a state issued inspection license before they can inspect automobiles. To obtain this license, technicians must take a safety inspection course and pass a written as well as a hands-on test before they can inspect automobiles.

In addition, Pennsylvania requires that automobile service facilities obtain a state issued license to sell Pennsylvania state safety inspections to customers. To become licensed, the service facility must show the state that it meets its requirements. These requirements include the ownership of required tools/equipment, minimum space, minimum insurance coverage to protect the customer's car, and a signed agreement to maintain records in accordance with state guidelines. Upon approval the service facility must display a state inspection sign with the station number on it (see Figure 1-9).

Similar to Pennsylvania's state safety inspection, the Environmental Protection Agency (EPA) requires regular automobile emissions inspections in some states. While state emission programs vary from state to state, they can be generalized as centralized or decentralized emission inspection programs.

FIGURE 1-9 A service facility with a Pennsylvania state inspection sign.

Centralized emission inspection programs require customers to drive their vehicle to a state-operated emission testing station to be inspected, as shown in Figure 1-10. If the automobile fails its emissions test, it cannot be repaired at the state-operated emission station but at an automobile service facility. State regulations may also require that the technician and/or the automobile service facility be licensed to perform the repair or re-inspection.

FIGURE 1-10 A centralized emission testing facility.

FIGURE 1-11 A decentralized state emission inspection station.

In some states the emission inspection program is decentralized (see Figure 1-11). This means an automobile service facility performs the emission inspection and the repairs an automobile needs to pass the inspection. Often, decentralized emission programs require that both the automobile technician and the automobile service facility have a license and own state-approved emission testing equipment.

When a state law requires that an automobile technician or automobile service facility have a license, it is typically issued by the state in which the facility does business. However, there are other federal EPA requirements to consider when working with, for example, automobile air-conditioning equipment. More specifically, automobile technicians must hold an EPA-approved license to reclaim, recycle, and recharge automobile air-conditioning systems. This certification license is required under the Clean Air Act Section 609 (see Figure 1-12). To obtain the federally mandated certification license to service automobile air conditioning, an automobile technician must obtain training and pass an EPA-approved test. This training can be obtained from companies that offer approved training courses.

At the same time, under the EPA's Clean Air Act Section 609 regulations, the automobile service facility does not need to be licensed. However, the automobile repair service facility must register with the EPA to assure them that it owns the required air-conditioning equipment that meets the standards set by the Society of Automotive Engineers (SAE).

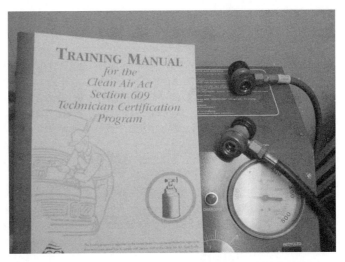

FIGURE 1-12 A training manual to obtain a Section 609 license to use the approved air conditioning recycling equipment shown.

ASE Certification

In some states, a license is not required for either automobile technicians or automobile service facilities. To show a level of competence many service facilities, regardless of whether they need a license, choose to participate in the voluntary certification of their employees (see Figure 1-13).

FIGURE 1-13 An ASE certification hanging on display at a service facility.

The most popular voluntary certification for automobile technicians as well as service consultants is offered by the **National Institute for Automotive Service Excellence (ASE)**.

The ASE requires technicians to document at least two years of work experience plus pass an exam to be certified within an automobile system. Specifically, there are eight ASE exams for each subsystem of the automobile, as follows: Engine Repair, Automatic Transmission/Transaxle, Manual Drive Train and Axles, Suspension and Steering, Brakes, Electrical/Electronic Systems, Heating and Air Conditioning, and Engine Performance. When technicians pass all eight exams, they are considered ASE master automobile technicians.

In addition to the ASE's examination for technicians, there are exams for specialty areas such as parts, engine machinist, compressed natural gas technician, and advanced engine performance certification, along with certifications in collision repair, heavy truck technician, and service consultant.

Another form of voluntary certification is provided by organizations such as AAA and A/C Delco. These organizations endorse automobile service facilities that meet certain standards.

Profit or Loss: The Bottom Line

After customers are served, they are charged money for the service. The money collected from a customer (not including any sales tax collected) is the **gross sale** amount. The total of customer gross sales is the gross sales for the automobile service facility.

As shown in Figure 1-14, the money collected from the customer (gross sale amount) pays the salaries of the automobile technicians and the cost of parts. The remaining balance is called **gross profit**. The gross profit is then used to pay the salaries of the other personnel, including the salaries of the managers and service consultant and the overhead expenses. Overhead expenses include rent, heat, light, telephone, uniforms, insurance, and the other expenses incurred when running a service facility. The balance left over after the expenses are deducted from the gross profit is called the **net profit** (see Figure 1-14).

The owner must pay the taxes of the service facility from the net profit. The balance is the profit after taxes and is the amount the owner may claim for a return on the investment of money and time spent working in the service facility. When the expenses cannot be paid by the gross profit, the negative balance is called the **net loss**. The money needed to cover a loss must come from the owner. If losses persist then the service facility will be bankrupt and must close.

GROSS LABOR and PARTS SALES
- <LESS> Cost of the technician's labor
- <LESS> Cost of parts to service the customer's automobile

EQUALS: GROSS PROFIT
- <LESS> Overhead expenses (examples)
  - ➤ Managers' salaries
  - ➤ Service consultants' wages
  - ➤ Rent (includes real estate taxes)
  - ➤ Heat
  - ➤ Electric
  - ➤ Insurance and benefits
  - ➤ Other expenses

EQUALS: NET PROFIT or LOSS
- <LESS> Taxes on the service facility

EQUALS: OWNER'S INCOME

FIGURE 1-14 The relationship between gross sales and profit.

Summary

Understanding the different types of automobile service facilities is important to service consultants. These differences may be recognized by the systems or products they service, such as a transmission system or a particular model of automobile.

Another difference among service facilities is their ownership. Service consultants must know who owns the facility and who is in charge of the operation. Is the facility a proprietorship, partnership, or corporation? Further, is it privately owned, a franchise, a chain, a dealership, or a fleet operation? In each case, the ownership and organizational diagram showing the chain of command has some special characteristics to be recognized by the service consultant.

Service consultants must also be knowledgeable about the state and federal licenses needed by a service facility. The legal requirements differ from state to state but must be met before some automotive services can be provided. In addition, employees should hold professional certifications, such as those of the ASE, because more and more customers look for a display that shows the personnel in the service facility are certified.

Finally, service consultants must be sensitive to the financial operations of their facility. In order to keep their facility healthy, service consultants must know the financial measures that go into the calculation of a profit or loss. As explained in this chapter, a service facility cannot remain in business unless a profit is earned.

Review Questions

Multiple Choice

1. An automotive repair business makes its money mainly by:
   A. selling parts to the customer
   B. selling the labor of a trained automotive technician to the customer
   C. selling the support staff's services to the customer
   D. selling the professional image of the repair facility to the customer

2. Whose job is it to work directly with the customer and technicians to arrange for diagnosis, repair, and service?
   A. The technician's
   B. The manager's
   C. The service consultant's
   D. The owner's

3. To "service" means to perform _____ on an automobile.
   A. maintenance
   B. repairs
   C. a diagnosis
   D. all of the above

4. According to the textbook, a profit business that services automobiles is called an:
   A. automotive shop
   B. automobile service facility
   C. automobile business
   D. automobile garage

5. An automobile service facility that repairs and maintains one system such as transmissions or brakes is called a/an:
   A. system-specific service facility
   B. product-specific service facility
   C. automobile dealership
   D. automobile garage

6. Which structure has stockholders?
   A. A corporation
   B. A proprietorship
   C. A partnership
   D. An automobile service facility

7. A _____ permits an owner to use a nationally recognized name and receive some assistance, such as training, advertising, and consulting, for startup operations.
   A. partnership
   B. proprietorship
   C. franchise
   D. corporation

8. A service facility that is one of several facilities owned by a corporation is called a:
A. chain service facility
B. specialty service facility
C. franchise
D. fleet service department

9. This text classifies a dealership service department as a type of:
A. chain service facility
B. specialty service facility
C. franchise
D. fleet service department

10. The _____ does not service customer vehicles but rather the vehicles owned by the company.
A. chain service facility
B. specialty service facility
C. franchise
D. fleet service department

Application Problems

1. Given the following information, calculate the parts sales of the service facility.

MONTHLY INCOME REPORT

GROSS SALES
    Labor $75,000
    Parts   ?
    Total Gross Sales $140,000

2. Given the following information, calculate the total gross sales, total cost of labor, and gross profit.

MONTHLY INCOME REPORT

GROSS SALES
    Labor $50,000
    Parts   40,000

    Total Gross Sales $_____

COST OF LABOR AND PARTS
    Technician Labor $25,000
    Parts   15,000

    Total Cost of Labor and Parts $_____
GROSS PROFIT $_____

3. Given the following information, calculate the net profit and the owner's income if the business taxes are 12.5%.

EXAMPLE OF FORM USED TO REPORT MONTHLY INCOME

GROSS SALES

Labor	$60,000	
Parts	60,000	
Total Gross Sales		$120,000

COST OF LABOR AND PARTS

Technician Labor	$32,000	
Parts	40,000	
Total Cost of Labor and Parts		72,000
GROSS PROFIT		$48,000

EXPENSES

Management Salaries	$11,600	
Service Consultants	8,600	
Rent	7,200	
Heat (oil)	1,400	
Electric	800	
Insurance and Benefits	3,000	
Other Expenses	3,400	
Total Expenses		36,000
NET PROFIT		$
Business Taxes		$ _____
OWNER INCOME		$

Short Answer Questions

1. Describe the different types of automobile service facilities.
2. Diagram an organizational structure for the different types of automobile service facilities.
3. Explain the importance of the guidelines that service facility operations must follow in different states.
4. Identify the major financial measures that have a direct impact on the profit or loss of a service facility.

CHAPTER 2

THE ROLE OF THE SERVICE CONSULTANT

OBJECTIVES

Upon reading this chapter, you should be able to:

- *Identify the major tasks of a service consultant.*

- *List the duties of a service consultant.*

- *Explain how the duties and tasks may be assigned in the different types of service facilities.*

- *Explain why the presentation of a professional image is important (Task A.1.14).*

Introduction

On a typical day a customer will enter an automobile service facility and determine within the first minute whether its professionalism and hospitality are acceptable. In turn, this impression will have an influence whether the customer thinks he or she will receive the service needed to solve his or her automobile concern.

To look professional, a service facility should be organized, clean, and proficient. This helps customers feel more comfortable and confident that the personnel at the service facility can solve their automotive problems. In addition, its employees (service consultant, technicians, and other support staff) must be interested and articulate when discussing problems and repairs with customers. When customers have a positive impression of the people working at a service facility, they are more likely to accept the service consultant's recommendations and have their automobile repaired there. Customers who are satisfied with a service facility's performance are more likely to become regular customers.

In order to produce satisfied customers, automobile service facilities must not only employ trained, experienced, and, in some states, licensed automotive technicians, they also need to have one or more professionally trained service consultants. The training ensures that the service consultants are positive, confident, and friendly and are able to put their customers at ease.

However, interacting with customers is only one part of a service consultant's job. The service consultant must perform a variety of other tasks to keep the work moving at the shop. This is important because a service facility must have enough customers flowing in and out to make a profit. This part of the service consultant's job can be complicated and therefore requires training and practice to become proficient. This proficiency cannot be gained overnight. To become proficient service consultants must first have a clear understanding of the tasks to be performed each day. They also must have knowledge of the different automobile systems serviced at their facility.

The purpose of this chapter is to provide an overview of the major tasks and duties of a service consultant and how they might differ among the various types of service facilities. In addition, because customer relations are often the key to the success of service consultants, this chapter gives special attention to their image (see Figure 2-1).

FIGURE 2-1 A service consultant at a customer station.

Major Tasks of Service Consultants

Job tasks are major work assignments given to employees and should appear in a job description. Each task has assigned **duties**, which describe the job requirements in detail the job requirements that an employee performs. Some jobs are quite complex, so a job task may have to be broken down into duties with related sub-duties. In this chapter the breakdown of the tasks is limited to duties and procedures.

ASE incorporates the tasks and duties of service consultants into the major skills and knowledge expectations of the position. These skills and knowledge areas are as follows:

- Communication skills
  - Customer relations
  - Internal relations
- Product knowledge
  - Engine systems
  - Drivetrain systems
  - Chassis systems
  - Body systems
  - Service/maintenance intervals
  - Warranties, service contracts, service bulletins, and campaigns/recalls
  - Vehicle identification
- Sales skills
- Shop operations

The tasks and duties discussed in this chapter focus on the operation of the shop. Communication and sales skills that are needed to accomplish the tasks and duties are discussed in the other chapters in this book.

Job Tasks of a Service Consultant

The owner or manager of the service facility ultimately determines the job tasks of the service consultant. Most of these tasks are common for all service consultants, although some owners may prefer to perform some of them, such as opening the business in the morning. Differences, of course, are usually found in the duties to be performed under a task and the related procedures as shown in the examples that follow.

First, the **job description** must list the tasks to be performed by the service consultant. The service consultant performs the duties to accomplish each task on a daily or weekly basis. An example of a job description for a service consultant is shown in Figure 2-2. This description was used

---

POSITION DESCRIPTION

Service Consultant

1. Opens shop at 7:30 a.m.
2. Greets customers, answers phone calls, provides information, makes appointments, calls customers for approval of work on vehicles, calls customers when jobs are completed, and places follow-up calls aftrer repairs are made.
3. Prepares customer invoices, reviews parts and labor charges with customers, and receives payment from customers (cash, check, or credit card).
4. Makes arrangements for customer shuttle or comfort if waiting at the service facility.
5. Prepares estimates, repair orders (RO), and computerized invoices (IN) and maintains the appointment book and customer status sheet.
6. Communicates with the technician to ensure service work is completed in a timely fashion.
7. Communicates with technician, as needed, to prepare estimates.
8. Assists, as needed, to help order parts, receive parts orders, and check parts invoices to ensure the charges are accurate.
9. Maintains inventory of office supplies, including supplies for computers, photocopier, fax, credit card machine, printers, forms, reports, and other materials, as needed.
10. Makes daily deposit to the bank and prepares daily report for management.
11. Closes the building at the end of the day and checks customer cars in the parking lot.
12. Other duties as assigned by management.

FIGURE 2-2 A job description for a service consultant.

in the repair facility owned by the authors, which is referred to as Renrag Auto Repair in this book.

At Renrag each employee was given a copy of his or her job description when hired. The tasks were reviewed with them before they started their job. When the employees' performances were evaluated, their performance on each job task was reviewed with them. If their job changed, then the job description was also revised and their pay was adjusted accordingly.

Note that Task 12, "Other duties as assigned by management," covers miscellaneous assignments that do not require enough time to make up a separate task. A good rule of thumb is that when a miscellaneous task takes more than 5 percent of an employee's workweek, it should become a separate task with assigned duties accompanied by a formal evaluation of the employee's performance of the task.

Finally, because Renrag employed an estimator/parts specialist, Task 8 required the service consultant to assist him. This task would not be in the job description if the service consultant calculated the estimates and ordered the parts.

Task Duties of a Service Consultant

Duties assigned to tasks are the details needed for service consultants to meet the expectations of a job.

This list of duties is helpful when training new employees and for evaluating their performance.

For example, at Renrag Auto Repair, the first task (see Figure 2-2) was to open the shop at 7:30 a.m. The duties for the first task are shown in Figure 2-3. These duties were listed according to rooms and bays with a list of specific functions to be performed. This detail provided for the opening of the facility and generated a checklist that was used by the service consultant or whoever opened the business.

Likewise, duties were outlined for the closing of the facility (see Figure 2-4). Of course, the owners could check the facility after it was closed to evaluate the service consultant's performances. If lights were left on or a thermostat was not set correctly, the owners could make a note and discuss it with the service consultant.

As shown in Figure 2-3 and Figure 2-4, the last duty for the opening and closing of the facility required the service consultant to work on the **customer automobile inventory sheet** (see Figure 2-5) to record the customers' automobiles that were left in the building and on the property. If the service consultant did not open the facility, the person who opened it, such as an owner, a manager, or a technician, used the customer automobile inventory sheet to check the automobiles left overnight or over the weekend. The customer automobile inventory sheet is important because losing track of a customer's vehicle is terribly embarrassing!

OPENING RENRAG AUTO REPAIR

FRONT OFFICE & WAITING ROOM
1. Enter front side door, turn on lights, and turn off security system.
2. Turn "CLOSED/OPEN" sign around to show "open."
3. Turn on lighted "OPEN" sign.
4. Unlock front door.
5. Turn on computer and printer.
6. Turn on TV security monitor.
7. Switch answering machine to position "A."
8. Turn on photocopy machine.
9. Turn on radio.
10. IF WINTER turn thermostat up to 72°.
11. Enter alignment bay.

ALIGNMENT BAY
1. Turn on lights.
2. IF WINTER turn thermostat up to 65°.
3. Turn on light in bathroom.
4. Go to inside bay.

INSIDE BAY
1. Turn on bay lights and lights to storeroom.
2. Enter compressor room.

COMPRESSOR ROOM
1. Turn on compressor.
2. Enter back bay.

BACK BAY
1. Turn on back bay lights.
2. IN SUMMER open bay door on the left (not facing Arch Street)
3. IN WINTER turn thermostat up to 65°.
4. Enter lube bay.

LUBE BAY
1. Turn on bay lights.
2. IN SUMMER open back bay door.
3. IN WINTER turn thermostat up to 70°.
4. Check for key drop envelopes.
5. Enter front office area.

INTERIOR OFFICE
1. Turn on lights.
2. Uncover computers.
3. Turn on computers and printers.

FRONT OFFICE
1. Listen to phone messages, check parking lot, and proceed to write up repair orders.

PARKING LOT—CUSTOMER VEHICLE INVENTORY
1. Use the **customer automobile inventory sheet** to check all customer cars left overnight in the building or in the lot.
2. Check customer automobiles left overnight for damages and missing parts not noted on the **customer automobile inventory sheet**.

FIGURE 2-3 Task 1 duties: Opening the shop at 7:30 a.m.

CLOSING RENRAG AUTO REPAIR

IF WINTER OR FRIDAY NIGHT—put all company automobiles, including the customer shuttle, in the back bay.

BACK BAY
1. IF WINTER turn thermostat down to 55° (if it is to go below freezing) or OFF if it is to stay above freezing.
2. Close bay doors (make sure motor turns off by looking at spindle in ceiling).
3. Make sure all droplights are off.
4. Check exit door (back wall) and make sure BAR is across the door.
5. Check water faucet (with hose) and make sure it is OFF.
6. Turn off bathroom light.
7. Turn off bay lights by compressor room.
8. Enter compressor room and turn off compressor.
9. Turn off compressor room light.

SIDE LUBE BAY
1. Go to lube bay and close and lock two bay doors.
2. Turn off tire balancing machine.
3. Make sure all droplights are off.
4. IF WINTER turn thermostat down to 55° if waste oil heater is on.
5. Turn off light to lube bay and upstairs light.
6. Go to internal bay lift and turn lights off.
7. Go to bay with alignment machine.

ALIGNMENT BAY
1. Close bay doors.
2. Make sure side exit door is closed.
3. Make sure all droplights are off.
4. Turn off alignment machine.
5. IF WINTER turn thermostat down to 55° (if it is to go below freezing) or OFF if it is to stay above freezing.
6. Turn off light in bathroom—leave door open.
7. Turn off bay lights.
8. Go into front office and enter interior office.

INTERIOR OFFICE
1. Turn off all computers and printers.
2. Cover computers with protective covers.
3. Make sure inspection sticker box is locked.
4. Turn of lights.

FIGURE 2-4 Task 11 duties: Closing the building.

CLOSING RENRAG AUTO REPAIR

OUTSIDE OFFICE & WAITING ROOM
1. Turn off all computers, printer, and TV security monitor.
2. Switch answering machine to position "B."
3. Turn off photocopy machine (leave fax ON).
4. Turn off radio and TV.
5. Turn off lighted OPEN sign.
6. Turn "CLOSED/OPEN" sign around to show "closed."
7. Make sure coffee pot is OFF.
8. IF WINTER turn thermostat down to 55° (if it is to go below freezing) or OFF if it is to stay above freezing.
9. IF SUMMER turn AC off.
10. Arrange furniture and magazines, put cups, etc. in trash.
11. Lock front door.
12. Set alarm, turn off lights, leave and lock door.

PARKING LOT—CUSTOMER VEHICLE INVENTORY
1. Check all vehicles in lot and make sure they are locked.
2. Record all customer automobiles left overnight and placed in the building and left in parking lot on the **customer automobile inventory sheet**.
3. Note any damages or missing parts on inventory sheet.
4. Enter date and time the inventory was taken.

FIGURE 2-4 Task 11 duties: Closing the building. [continued]

RENRAG AUTO REPAIR

Customer Automobile Inventory Sheet

Date & Time	Year	Customer Automobile Make Model (note location and condition)	License Number	RO Number

FIGURE 2-5 Customer Automobile Inventory Sheet.

Use of Duties to Set Procedures

At Renrag Auto Repair the duties for certain employee tasks overlapped. Therefore the tasks were combined to create a set of procedures employees could follow. For example, the procedures to process repair orders for customer repairs included interaction between the service consultant, parts specialist, and technician for Tasks 2 through 8, shown in Figure 2-6.

Note: Service Consult (SC), Estimator/Parts Specialist (ES), and Technicians
1. Service Consultant (SC)
   a. Greets customer—obtains information on customer name, phone number, address (if necessary), vehicle repair to be performed (**complaint**), time of pick-up, and enters this information on the computer repair order (RO).
   b. Enters job into computer by customer name and computer assigns a number to the repair order (RO).
   c. RO is printed and customer signs copy.
   d. RO is entered on the SC's **status sheet** (see Figure 2-7).
   e. RO is placed on technician's wall—if the work is a maintenance service or a state inspection, the SC includes the appropriate form on the clipboard with RO.
2. Technician
   a. Performs the diagnosis, repair, or maintenance.
   b. If a diagnosis, the technician writes the repair to be made (**cure**), the parts needed, and the reason (**cause**) for the repair on the back of the RO.
   c. The RO is returned to the SC.
3. SC
   a. Reviews the diagnosis (if repair is made or service completed, go to step 9).
   b. Makes the notation on the SC **status sheet**.
   c. Forwards RO to the Estimator/Parts Specialist (ES).
4. Estimator/Parts Specialist (ES)
   a. Prepares parts estimate and records BOTH the cost of the part and the price to charge the customer on an estimate form.
   b. Prepares the labor estimate of hours and cost of each repair to be made on the vehicle on the estimate form.
   c. Reviews the estimate with the technician.
   d. Returns RO with the estimate to the SC.
5. SC
   a. Updates the SC **status sheet** to indicate estimate has been completed.
   b. Calls customer with estimated cost of the repair for approval.
   c. *If the customer does not approve* the repair, go to step 9.
   d. *If the customer gives verbal approval for* the repair, the SC:
      i. enters the approval on the SC **status sheet.**
      ii. informs the technician.
      iii. returns the RO to the Estimator to order the parts.
6. ES
   a. Places the order for the parts.
   b. Informs SC that parts have been ordered in order for the SC to make entry on the SC **status sheet**.
   c. Receives the parts and:
      i. checks the part and the delivery receipt or vendor invoice to ensure the part is correct item and the cost on the invoice and the cost on the estimate form are the same.
   d. Enters the part number, name of part, and cost on the front of the RO.
   e. Makes photocopy of delivery receipt and places the copy with the RO.
   f. Delivers part to technician.
   g. After all parts for a job have been received,
      i. writes the labor to be charged on the front of the RO.
      ii. gives the RO, job estimate form, and copies of the delivery receipt or vendor invoice to SC.

FIGURE 2-6 Processing repair orders.

7. SC
   a. Updates SC **status sheet**.
   b. Checks that the information on the RO and the job estimate for parts and labor are the same (if not, then the differences need to be resolved).
   c. Forwards RO to technician.
   d. Enters the parts and labor information on the computer RO.
8. Technician
   a. Makes repairs.
   b. Test drives the vehicle and documents on RO.
   c. Returns RO to SC.
9. SC
   a. Updates SC **status sheet** indicating job is complete.
   b. Completes computer RO (includes tech comments and recommendations if appropriate).
   c. Prints invoice for customer to pay (IN).
   d. Staples RO, copy of delivery receipts or vendor invoices and estimate sheet to the hardcopy of the IN, and places IN in the upper tray.
   e. Informs customer that job is complete and indicates on the **status sheet** that the call was made and the time.
   f. When customer arrives, reviews IN with customer, receives payment, indicates method of payment on IN, gives customer the keys to the vehicle, files IN in lower tray, and records pick-up & time on SC **status sheet**.

FIGURE 2-6 Processing repair orders. [continued]

In the process presented in Figure 2-6, a key form used by the service consultant was the customer status sheet (see Figure 2-7). This form is extremely useful when tracking the automobiles being processed through a facility, especially on busy days when there is a large number of automobiles being checked in and picked up. As each step in the process is com-

Renrag Auto Repair

Customer Status Sheet

Fill in information or check when completed
(write in N/A in box if not applicable)
(enter date and time of pick-up)

RO No.	Cust. Name	Model & Yr. Auto	Cust. Sign RO	Diagnosis Done	Est. Complete	Cust. Approved	Parts Ordered	Parts Rec'd	Job Completed	Cust. Called	Picked Up and Paid

FIGURE 2-7 A service consultant's status sheet for customers.

pleted, the service consultant should record the time in the box provided. When a customer calls about the status of a repair on his or her automobile, the person answering the phone can quickly give an update by looking at the status sheet to see where the automobile is in the process. If an automobile has been at one step for an unusually long time, the status sheet can be examined to determine where a potential problem is occurring.

In addition to the status sheet and inventory forms, other forms were also used by the service consultant at Renrag Auto Repair. For example, Task 10 (Figure 2-2) required the use of a special form to establish an accounting control. Specifically, Task 10 required the service consultant to make the bank deposit and prepare a report for the owners. The form in Figure 2-8 was used to perform this task and related duties. The informa-

---

RENRAG AUTO REPAIR, INC
Receipts for the Day

Date: _____

Cash:		Checks:		Credit Cards	
IN#	Payment	IN#	Payment	IN#	Payment

Total Receipts $ _____   Bank Deposit $ _____

Total Cash $ _____

Total Checks $ _____   Payment for work done on credit:
                                          Name   IN#   PMT

Total Cr. Cards $ _____

Total Receipts $ _____

                                          Work completed today on credit:
                                          Customer   Amount
                                          Name   IN#   Owed

NO. of ROs   _____

NO. of LOFs _____

FIGURE 2-8 Task 10: Making daily deposits and preparing daily reports for owners.

tion recorded daily on this form was used to provide a picture of the daily, weekly, and monthly operations. For example, the information recorded on the number of LOFs per day was graphed, giving a picture of the daily, weekly, and monthly volume of oil changes sold by the business. This information was used to change prices and offer specials on the least busy days of the week.

Duties at Different Types of Service Facilities

Different types of service facilities have different tasks, which are important, and possibly use different forms to record processes and daily activities. For example, specialty service facilities such as those that perform a transmission service do not have to employ service consultants with extensive product knowledge in other systems, for example, chassis systems. Likewise, service consultants in fleet service departments may not have to have the extensive communication skills needed for customer relations or sales. In both cases, the forms used are different from those of independent facilities that service all systems on all makes and models.

Because franchise and chain service facilities have predetermined procedures with forms and possibly written scripts for service consultants to follow, the duties for certain tasks may differ from those of independently owned facilities. For example, to ensure that service consultants at franchise service facilities understand their duties, they often attend training prior to being hired.

Dealership service departments may also require extensive training for service consultants. The manager, an outside consultant, or the automobile manufacturer may conduct this training. The focus of the training depends on who conducts it. For example, automobile manufacturer training typically focuses on warranty administration, while many outside consultants focus on sales skills and communications (see Figure 2-9). Service managers who conduct internal training sessions often focus on shop operation, communication among employees, and procedures.

Finally, the larger corporate and multi-manufacturer auto dealership repair facilities may have specialists who can relieve the service consultant of some of the tasks typically performed at smaller facilities. In the job description shown in Figure 2-6, the service consultant was relieved of some tasks and duties by the parts specialist. This was not the case when Renrag was a new facility and had fewer customers. Also, at a dealership a cashier usually receives customer payments instead of the service consultant. The duties assigned to a service consultant must therefore reflect the tasks needed to ensure an effective and efficient daily operation. If a service facility grows or adds products, the job descriptions must be changed.

FIGURE 2-9 A service consultant with a customer.

Professional Image

An automobile service facility must present a professional image, and the service consultant must be able to convince customers that the service facility can and wants to help them. The service consultant and other support staff must recognize the customers' presence by smiling and cheerfully saying "Hello" even when the automobile service facility is very busy. If customers are recognized and made to feel comfortable, they are more likely to be patient and wait until they can be helped.

Furthermore, when the service consultant is trained to follow a sequence of steps and project an attitude that conveys interest in solving the customers' problems, they will be more at ease, begin to feel "at home," and be more receptive to the suggestions and recommendations of the service consultant and technicians.

Customers also notice the personal appearance of the service consultant and other employees as well as the appearance of the facility. Appearances often give customers a feeling about the care a facility will give to their automobile. To convey this attitude, a service consultant should be clean and neatly dressed. While some odors associated with an automobile, such as oil and gas, cannot be avoided, employees must be sensitive to the odors related to their personal hygiene (some experienced service consultants keep a bottle of mouthwash, deodorant, and aftershave lotion at their station). Many service consultants and other employees often wear a uniform or shirt with a name or nametag. Likewise, the furnishing and cleanliness of the customer reception area should be constantly monitored (see Figure 2-10). If a facility does not care about the appearance of its employees or business, it implies that the care of a customer's automobile is also not important!

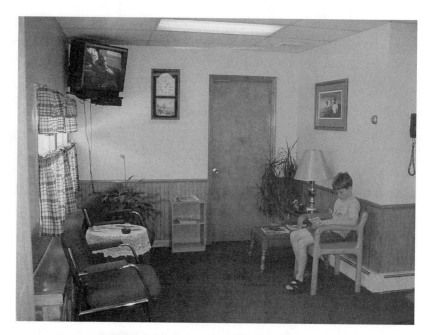

FIGURE 2-10 A neat and clean waiting area.

Therefore, a professional image and the right attitude are a fundamental key to the success of a service facility. The service consultant must remember that the service facility is in business to serve the customer. This includes every employee in the service facility, not just the service consultant. Everyone must take the time and be willing to resolve the customers' automotive problems.

Summary

Today's busy automobile service facilities present a challenge for service consultants. As shown in the chapter, the various activities require them to manage multiple tasks at the same time. For example, daily activities such as *answering the phone*, talking with technicians, *answering the phone*, writing estimates, *answering the phone*, providing updates to customers, *answering the phone*, finalizing invoices, *answering the phone*, and interacting with customers who have just arrived are all taking place in the same time frame. Of course, the point is that the phone can be a great interruption, but it also is the connection to customers. This means that the service consultant must be able to be able to multitask. For some people this may be a serious problem!

The complex job of a service consultant must be designed to fit the facility. This can only occur if the job description is carefully prepared and clear and the service consultant is given sufficient details through related duties. Next, the job descriptions of the service consultant and other employees must fit into the process and procedures of the facility. When this exists, the service consultant can be organized, can think ahead, be able to serve customers, and be able to assist the technicians when necessary.

Finally, the service consultants' conduct must be professional and be very accurate in what they sell to customers. They must never promise to do more than required or possible, which is called an "over promise." In addition, service consultants must be sure that repair estimates are accurate and a profit is earned on each repair. Without a profit on each job and a steady stream of new and repeat customers, service facilities cannot remain in business and solvent.

Review Questions

Multiple Choice

1. A service facility must have enough customers flowing in and out to make a:
   A. sale
   B. business
   C. technician satisfied
   D. profit

2. What organization incorporates the tasks and duties of service consultants into a list of the major skills and knowledge expected of the position?
   A. The Automobile Manufacturers Association
   B. The Service Facility Association
   C. The Automotive Service Excellence
   D. The American Automobile Association

3. Who ultimately determines the job tasks of the service consultant?
   A. The service consultant
   B. ASE
   C. The owner or manager
   D. The technician

4. What form is extremely useful when tracking the automobiles being processed through a facility, especially on busy days when there is a large number of automobiles being checked in and picked up?
   A. Status sheet
   B. Customer Automobile Inventory Sheet
   C. Opening RENRAG Auto Repair Sheet
   D. Job description for a service consultant

5. Customer service can be enhanced by:
   A. a clean and organized facility
   B. the willingness of all employees to help customers
   C. eliminating the smell of gas and oil from the shop area
   D. A and B
6. Which of these can have the greatest impact on a customer's decision to do business with you?
   A. Extended business hours
   B. The service consultant's appearance
   C. The level of trust they feel
   D. Discount pricing

Application Problems

RO No.	Cust. Name	Model & Yr. Auto	Cust. Sign RO	Diagnosis Done	Est. Complete	Cust. Approved	Parts Ordered	Parts Rec'd	Job Completed	Cust. Called	Picked Up and Paid
100	Jones	88 Tempo	X	X	X						
101	Thomas	02 F150	X								
102	Grant	87 Lancer	X	X	X	X	X				
103	Meyers	04 Camry	X	X	X	X	X	X			
104	Sloan	04 Camry	X	X	X	X	X	X	X	X	

1. Which automobile is waiting for parts to arrive?
2. Which automobile is waiting to be repaired with parts that have just arrived?
3. Which automobile has not yet been examined by a technician?
4. Which automobile has been fixed and the customer called?
5. Which automobile is waiting for customer approval?

Short Answer Questions

1. Identify the major tasks of a service consultant.
2. List the duties of a service consultant.
3. Explain how the duties and tasks may be assigned in the different types of service facilities.
4. Explain why the presentation of a professional image is important (Task A.1.14).

CHAPTER 3

THE TEAM APPROACH

OBJECTIVES

Upon reading this chapter, you should be able to:

- *Explain why and how a team approach can offer superior customer service.*

- *Present optional team formations and team member assignments in different types of service facility operations.*

- *Identify the major responsibilities of the team leader.*

Introduction

In the 1950s automobiles had very basic mechanical systems when compared with today's systems, which are mostly computer controlled. For example, air conditioning was not popular then and the only electronic system was the AM radio. Systems such as antilock brakes were not even considered possible. In addition, the makes and models of automobiles were essentially limited to the big three (Ford, GM, and Chrysler) with only a few other makes and models that were mostly American but did not last, such as Packard, Hudson, and Studebaker. The operations of the various automobile systems were similar and repair manuals were rarely needed. What is now known as an automobile service facility was referred to as a repair garage, and service was generally quite different when compared with today's facility.

One of these repair garages in the "good old days" was owned and run by John. His was an independent repair shop and did not pump gas like many repair garages. John did not talk much and worked alone, except when he could convince his sons to help him. People took their cars to John because of his mechanical ability. He had what was known as a "good ear," meaning he could tune an engine by listening to it run. He was considered honest because he did not put used parts on the cars he repaired and then charge for new parts. His repair charges were not cheap but he was considered reasonable.

John liked his one-man shop, which meant he was the service consultant, parts specialist, technician, bookkeeper, receptionist, janitor, and groundskeeper, unless he could get one of the boys to mow the grass. When customers wanted John to repair their car, it was expected that they would go into the shop, find John, and wait patiently until he wanted to pay attention. The customers typically had to explain the problem to John while he continued to work on something else. After John agreed to fix the car, the customers would leave it out front and then return, possibly several times, to see if it was done. Calling John did not work very well because he often did not hear the phone when he was working on a car. Eventually, customers would discover that their car was out front again, which meant the repair was finished. John only liked to fix one to five cars a day so he could do bookkeeping at night.

Obviously John's Garage could not survive in today's market without some of the modern service facility improvements. For example, aside from greater business expenses, such as heat, light, uniforms, advertising, insurance, and benefits, there are technical advances that require expensive equipment. In addition, customer service has changed. Customers expect personal service and professional treatment when they take their automobile to a service facility. This, of course, can be taken as a compliment because the image of the repair garage in John's day has shifted to a service facility with trained professionals. Customers expect to be treated differently by professionals who are trained to perform a specific job as compared to John's day.

The tool investment experiences of professionals at a service facility have changed also. John's tools were a basic set of hand tools, a droplight, a jack, and a creeper. This meant that John's personal investment was not very large. In addition, he had few business expenses outside of his parts bill. Therefore, servicing a large number of customers was not necessary so two service bays were more than enough. To make a profit, John usually worked six days a week and all invoices were paid by cash or check (credit cards were not in use yet), which made business a lot easier. Today, the investment in setting up and running a service facility is very different. The costs for tools, equipment, and technicians with high levels of expertise have increased dramatically. A fully equipped facility needs multiple bays and a convenient, sometimes expensive, location to conduct the business required to make a profit. Furthermore, there is also the expectation of personal and professional service.

More specifically, in order to earn the money needed to meet the costs of an automobile service facility and remain competitive with respect to the amount charged to customers, a facility must process a set number of customers over a "five-day workweek" to break even. To serve the necessary number of customers per day to make a profit and provide the personal service expected by customers, a facility may need to employ people to perform different jobs, such as a service consultant, a parts specialist, a driver for a customer shuttle, and others. However, just having these different types of employees on the payroll is not enough. They must know their job and be able to perform it effectively. To be efficient, employees must also work together as a team. Certain people can be the best at performing a job, but their inability to work with others could mean they are not useful to the business.

The purpose of this chapter is to discuss the composition and use of teams at service facilities and how these teams may differ from one facility to another. Additionally, the chapter explains the importance of leadership. Teams will not function effectively unless someone is willing and able to provide the leadership necessary to get them to work together every day.

Purpose of the Team Approach

Like John's Garage, the use of teams at service facilities has changed in recent years. More specifically, a service facility where the technicians worked together as a team is no longer adequate for the larger, more complex facilities. To attract the more desirable customers, greater efficiency with expanded services and faster delivery are required. This calls for an approach whereby everyone in the facility is on a team and works together. When customers enter a facility, the services they receive should proceed through a seamless system, beginning with an introduction by the service consultant and ending when they are handed the keys to their automobile.

Basically, customers must be able to enter a facility with minimal effort and confusion and have their problems solved in a timely manner. Although their primary objective is to have the problem with their automobile solved, there are a number of other desired services to be recognized, such as:

- transportation to home or a job while their automobile is at the facility
- a comfortable and pleasant place to wait while their automobile is in the shop
- a phone to call someone to let him or her know where they are or to leave a message requesting transportation
- assistance in obtaining credit for the payment of a large repair
- quick, and sometimes confidential, service when paying a bill
- follow-up appointments for service and repairs
- information about the different quality or grades of parts available, such as tires
- restroom facilities and drinking water
- a referral to a specialty business if necessary
- updates on the progress of their service

Obviously, one person cannot provide all of these services to the customers.

Two Primary Missions of a Team

The members of a team cannot just do their job; they must relate it to the team's primary missions. The first is to work together to ensure their customers are treated as important people and are satisfied with the service and treatment they receive. Satisfaction in this case is seen as a responsibility of all of the members on the team.

In addition to customer satisfaction with effective treatment, the efficiency of the facility is the second mission. This is important because it is related to production. Greater efficiency leads to greater production. Efficiency is also related to customer satisfaction. Efficient service is one of the important expectations of customers.

In order to meet the costs generated by a modern service facility, it must earn the money needed to pay its bills. If not enough money is earned, the efficiency of the operations may be the reason. Too often owners and service facility managers think that having more service bays will generate more money. This may not be correct and may actually drive up the costs and lower the average number of billable hours per bay per day. One method to calculate billable hours per day per bay is to take the total number of dollars earned by the service facility for labor divided by the service facility's labor rate per hour charged to customers divided by the number of bays. For example, $4,000 of labor charges per day ÷ $80 per hour ÷ 10 bays = 5 hours per bay per day.

As a result, the objective for production is to maximize the number of hours of service generated in each service bay. For example, if the bays in a facility are generating an average of four to five billable hours of service per day, and the facility is not profitable, it will have to increase the number of billable hours to perhaps six or seven hours per day to stay in business. To do this, the facility will have to either increase the number of customers served per day or the number of hours charged to each customer.

To increase productivity by either method may require employees to work together more efficiently, not necessarily "harder." One method to improve efficiency is to implement the team approach in order for employees to work together effectively. If a team already exists, it does not mean that its members are working together efficiently or effectively. So if production is not adequate, it is possible that the members do not understand the purpose of the team, or they may need more direct leadership.

At some service facilities, an efficient and hardworking team may already exist but may need to be expanded. For example, when the service consultant is too busy answering the phone or collecting money from customers, jobs may not be moving through the system fast enough. In such a case, additional personnel may be needed to permit the service consultant to focus on getting more customers' automobiles into and through the system.

The Team Process

The team approach was used by Renrag Auto Repair to process the repair orders described in Chapter 2 and in Figure 2-6. In order to discuss this team approach, the process is shown again in Figure 3-1. The difference between the two figures is the words that are in bold print. In Figure 3-1 the actions taken with the repair order (RO) are in bold print. This is to point out how important the RO is to the process. Team members must learn to "keep their eyes on the ball," which, in this case, is the RO.

The team at Renrag had three members: the service consultant, the estimator/parts specialist, and the technician. Each team member had to be knowledgeable about the entire process and know when the ball, which was the RO, had to be passed to the next member on the team. The handoff had to occur promptly after the team member completed all of the assigned duties.

As shown in Figure 3-1, as soon as the service consultant completed an RO, it was placed on a peg for the technician. The service consultant could not delay because the entire process would be held up and work would back up. Likewise in step 4 the parts specialist had to return the RO with the estimate to the service consultant. Again, this had to be given to

Note: Service Consult (SC), Estimator/Parts Specialist (ES), and Technicians
1. Service Consultant (SC)
   a. Greets customer—obtains information on customer name, phone number, address (if necessary), vehicle repair to be performed (complaint), time of pickup, and enters this information on the computer repair order (RO).
   b. Enters job into computer by customer name and computer assigns a number to the repair order (RO).
   c. **RO is printed and customer signs copy**.
   d. RO is entered on the SC's status sheet (see Figure 2-7).
   e. **RO is placed on technician's wall**—if the work is a maintenance service or a state inspection, the SC includes the appropriate form on the clipboard with RO.
2. Technician
   a. Performs the diagnosis, repair, or maintenance.
   b. If a diagnosis, the technician writes the repair to be made (cure), the parts needed, and the reason (cause) for the repair on the back of the RO.
   c. **The RO is returned to the SC.**
3. SC
   a. Reviews the diagnosis (if repair is made or service completed, go to step 9).
   b. Makes the notation on the SC status sheet.
   c. **Forwards RO to the Estimator/Parts Specialist (ES)**.
4. Estimator/Parts Specialist (ES)
   a. Prepares parts estimate and records BOTH the cost of the part and the price to charge the customer on an estimate form.
   b. Prepares the labor estimate of hours and cost of each repair to be made on the vehicle on the estimate form.
   c. Reviews the estimate with the technician.
   d. **Returns RO with the estimate to the SC.**
5. SC
   a. Updates the SC status sheet to indicate estimate has been completed.
   b. Calls customer with estimated cost of the repair for approval.
   c. *If the customer does not approve* the repair, go to step 9.
   d. *If the customer gives verbal approval for* the repair, the SC:
      i. enters the approval on the SC status sheet.
      ii. informs the technician.
      iii. **returns the RO to the Estimator to order the parts.**
6. ES
   a. Places the order for the parts.
   b. Informs SC that parts have been ordered in order for the SC to make entry on the SC status sheet.
   c. Receives the parts and:
      i. checks the part and the delivery receipt or vendor invoice to ensure the part is correct item and the cost on the invoice and the cost on the estimate form are the same.
   d. Enters the part number, name of part, and cost on the front of the RO.
   e. Makes photocopy of delivery receipt and places the copy with the RO.
   f. Delivers part to technician.
   g. After all parts for a job have been received,
      i. writes the labor to be charged on the front of the RO.
      ii. **gives the RO, job estimate form, and copies of delivery receipt or vendor invoice to SC.**

FIGURE 3-1 Processing repair orders.

7. SC
   a. Updates SC status sheet.
   b. Checks that the information on the RO and the job estimate for parts and labor are the same (if not, then the differences need to be resolved).
   c. **Forwards RO to technician**.
   d. Enters the parts and labor information on the computer RO.
8. Technician
   a. Makes repairs.
   b. Test drives the vehicle and documents on RO.
   c. **Returns RO to SC**.
9. SC
   a. Updates SC status sheet indicating job is complete.
   b. **Completes computer RO** (includes tech comments and recommendations if appropriate).
   c. Prints invoice for customer to pay (IN).
   d. **Staples RO**, copy of delivery receipts or vendor invoices and estimate sheet **to the hardcopy of the IN**, and places IN in the upper tray.
   e. Informs customer that job is complete and indicates on the status sheet that the call was made and the time.
   f. When customer arrives, reviews IN with customer, receives payment, indicates method of payment on IN, gives customer the keys to the vehicle, files IN in lower tray, and records pickup & time on SC status sheet.

FIGURE 3-1 Processing repair orders. [continued]

the service consultant promptly so that the customer could approve the job. If the customer was waiting in the facility's waiting room, steps 2, 3, and 4 had to be given priority and all team members had to be informed of the priority by the service consultant.

In the process shown in Figure 3-1, the service consultant is the pivot and is expected to monitor the processing of the RO. However, all members are expected to "keep their eyes on the ROs" in order to keep the work moving efficiently through the facility. In the process the service consultant plays a major role at the beginning (step 1, initial customer contact), middle (step 5, customer approval to perform the work), and end of the process (step 9, receives payment from the customer for the repair). Consequently, because of their duties and position on the team, service consultants may be considered the team leader by default.

A Seamless System

The handoff process should be a seamless system, meaning it should not be disrupted by having to move to another process and then return at some point to be continued. In Figure 3-1 the only external delay is if there is a need to obtain repair parts. A flowchart for the system is shown in Figure 3-2. In the flowchart, each block corresponds to the process shown in Figure 3-1.

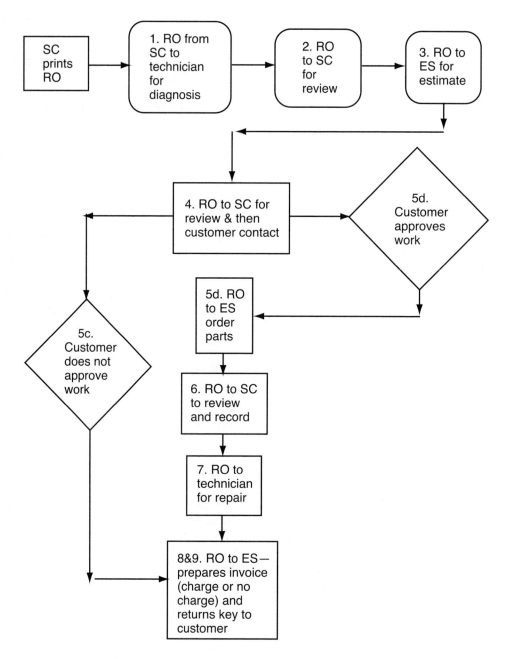

FIGURE 3-2 A flowchart for the system to process an RO.

Modifying the Team and System

At Renrag, if the owners had continued to run the business and the volume of customers had continued to increase, the number of members on a team would have had to be expanded or a second team created. The other support personnel that may have been added to the team would have been:

- a cashier to receive payments and make the evening deposits. The payment in step 9 would be given to a person in the business office (not the service consultant).
- a receptionist to answer the phone and make contact with the customers in steps 5 and 9.
- a parts runner to take parts to technicians and go for parts and supplies when needed. Step 6 would include this employee in the process.
- a customer shuttle driver
- an employee to clean customers' cars after a repair. A step would have to be added between steps 8 and 9.

Flow Diagrams on Working Relationships of Team Members

A **flow diagram** is often helpful for team members so that they can "see" how they are to work with each other. A diagram illustrating the working relationships at Renrag Auto Repair is shown in Figure 3-3. As this diagram exhibits, the owners (or a service manager) were responsible

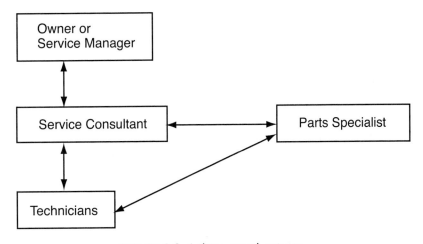

FIGURE 3-3 A three-member team.

for the team performance and all of the members on the team interacted with each other. The process shown in Figure 3-1 represents this working arrangement.

To ensure that the process was working, the owners at Renrag met with the team on a weekly basis either before the shop opened or at the end of the day. In addition, on days when either a large number of customers had appointments or a large repair job was taking several days to complete, the owners assisted the service consultant, parts/estimator specialist, or technicians.

Adding Team Members to a Diagram

From experience, the owners of Renrag knew that if their six-bay garage was not processing twenty to twenty-five customer vehicles a day, if the appointment book was filled for more than two days, if the bays were not generating at least 6 billable hours per day, (every team member was working at maximum capacity) a bottleneck existed somewhere. If investigations found that more technicians were not needed, then the suspicion was that the service consultant and parts specialist needed some assistance to increase the flow of customers into, through, and out of the facility. In this case, the owners (or a service manager) should provide the initial assistance to determine whether or not assistance opens up the bottleneck. If the assistance is successful, then a support staff person may be justified.

When support staff is added to a team, they must be prepared to perform their job and must understand the team concept. This seemingly minor point cannot be stressed enough. When a part-time or full-time staff member is hired, he or she may not be able to support a team as expected initially, and it may take time before any noticeable change occurs in productivity.

For example, assume that a cashier is added to the team in Figure 3-3 and a technician is assigned to be the lead technician to help assign work to other technicians. If this is the case, a new diagram (see Figure 3-4) needs to be created to show their working relationships. Notice in Figure 3-4 that the service consultant and parts specialist have a working relationship on a daily basis with the cashier and lead technician.

The Role of a Lead Technician

In Figure 3-4, all work flows through the **lead technician**, who assigns jobs to the other technicians. When a service facility has a lead technician, the team leader responsibilities are shared between the service consultant and lead technician. The service consultant is technically a part of management and, in that capacity, works with the service manager on issues related to the team's customer relations, warranties, the processing of repair orders, general operations, sales, volume of work processed,

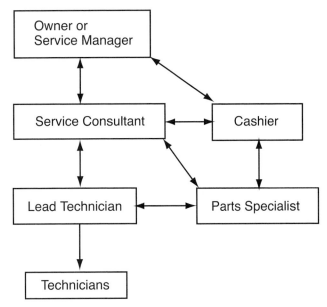

FIGURE 3-4 A four-member team.

and so on. At the same time, the lead technician assists with coordination of the work within the shop, handing out work to each technician on his or her team, shop operations, quality control, monitoring the condition and use of shop equipment, and so on.

The lead technician is expected to work as a technician in addition to handing out work to other technicians. This is important because jobs must be handed out both to the technician who can do the job and in a manner that is fair to all technicians. Because of the problems caused by the way jobs are assigned, some facilities have the technicians elect the lead technician. As a result, the lead technician may not be the service facility's best technician but someone the technicians respect and trust to assign the work fairly.

The assignment of jobs by the lead technician is critical when the technicians are paid by the job in a system called **flat rate**. The flat-rate system gives a technician a certain amount of time to do a job. For instance, a job may be assigned a flat rate time of 1.5 hours. If the technician completes the job in 1.2 hours, he or she is still paid for 1.5 hours of work. The technicians' pay at the end of the week is based on the total amount of time they are given to complete the jobs assigned (not the time it actually takes to complete the jobs) multiplied by their pay rate.

For example, if a technician earned 50 flat-rate hours for the week, possibly taking 38.8 hours to complete all of the jobs, and their pay rate is $10 per flat-rate hour, then the pay would be $500. Therefore, for technicians who are paid by flat rate, how work assigned is important. If a job takes longer than the time allowed under the flat rate, then the technician will lose time and, ultimately, money. Of course, on occasion, technicians expect to lose time on some jobs. However, when technicians are constantly assigned work that loses money, they will have a morale problem and may even quit their job. Consequently, assigning work fairly to technicians capable of doing the work is critical to the effective and efficient performance of a team.

Creation of Multiple Teams

When the volume of customers at a facility increases to a level in which the bays are consistently producing an above average number of billable (flat-rate) hours per day, it may need to create an additional team. After the additional team is set up, the facility should begin to process more customers per day and increase its total number of billable hours. At the same time, the facility should also show a decrease in the average number of billable hours per bay to the desired average number of hours. This expansion depends on several factors, such as space for more bays and equipment, money to buy the equipment, and availability of technicians and staff to hire.

Figure 3-5 shows a diagram of the working relationship for two teams. In this diagram the support staff, such as the cashier, works for the service manager, the parts specialist works with both teams, and the technicians have been split to work with one of two lead technicians. If the volume increases to the point where the parts specialist cannot handle the work, a second specialist may be employed to work with each team.

In some cases, the demand for specific services may increase, indicating a need to modify a team's composition. As a result, other options may be considered. For example, at Renrag there was a growing demand for the maintenance of automobiles, specifically quick oil changes, tire rotations, antifreeze fluid changes, and air-conditioning inspections and maintenance, among others. Because of the demand, one option would have been to create a **specialty team** specializing in automobile maintenance while the other team continued to do the diagnosis and repair of automobiles. This arrangement and working relationships would have been similar to the ones shown in Figure 3-5.

Naturally, the addition of a specialty team requires the process to be carefully changed because, in some cases, a customer's automobile will be serviced by both teams. For example, one team may change the oil on the automobile while the other may make brake or engine repairs. The arrangement whereby personnel, technicians, and service bays are equipped for the jobs they perform promises to be very efficient. In other words, within the general service facility would be a specialty service facility.

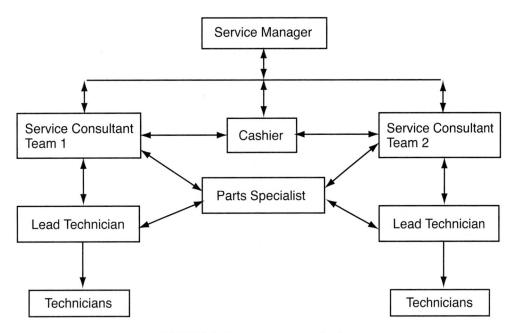

FIGURE 3-5 Two teams in one facility.

Teams at Different Types of Facilities

In reality, a small independent service facility's team is comprised of a service manager, who also serves as the service consultant. In many cases, the service consultant also orders parts and prepares the estimates. If the owner or service manager/consultant has a technical background, he or she may also serve as a technician on occasion. However when in the technician role, this person should only "pick up the slack" to help the busy technicians. Otherwise the service manager/consultant should oversee and hand out work to the technicians. The relationship structure of the service facility would therefore resemble the diagram shown in Figure 3-3. It is also not uncommon for a part-time or full-time staff assistant to be hired at these facilities to answer the phone, call for parts, receive deliveries, file documents, and so on.

At larger service facilities, more "support staff" are often needed. The diagrams for these working relationships would resemble the one shown in Figure 3-4. Naturally, as more technicians are hired, more service consultants and parts specialists are needed. A good rule to follow is that, typically, one service consultant can handle between four and six technicians. Important details, such as returning a customer's call, are likely to be missed beyond this number. When additional service consultants are hired, the diagram for the working relationships would resemble Figure 3-5.

To make a larger dealership service department more manageable, the service facility may be divided into smaller work units called lateral support teams. This means a service manager oversees several teams made up of a service consultant, a lead technician, and three to five technicians. The structure is similar to that of the larger service facility, except it uses one of the technicians as a team leader. The working relationship diagram would resemble Figure 3-4.

At some dealerships the teams are referred to as a group. The technical difference between a team and a group system is how the technicians are paid, not the structure itself. The team system typically averages the technicians flat rate hours and the pay for each technician is based on the team's average. Some feel this produces better cooperation while others point out that highly efficient technicians are penalized. The group system typically pays each technician the flat rate hours earned regardless of the other technicians' performance.

When a lead technician is not used in a team arrangement and the service consultant is too busy to assign work to the technicians, a "dispatcher" may be added to the diagram. This person is typically responsible for assigning work to the technicians and reports directly to the service consultant. This arrangement is more common at some chain store-owned service facilities, franchises, and smaller dealerships where there are not enough technicians to form separate teams but there are too many for the service consultant to oversee.

Finally, at some automobile dealership service departments and many chain store-owned service facilities, it is not uncommon to have more formal working relationships between the technical teams and the nontechnical support staff team. The nontechnical support staff may include service consultants, parts specialists, an assistant to the service manager, a cashier, a warranty clerk, bookkeepers, custodians, automobile detail and lot personnel, and a facility maintenance crew. A formal working relationship between the nontechnical support staff and the technical personnel in the service facility may resemble the diagram shown in Figure 3-6. As shown in this diagram, although the technical and nontechnical people may communicate informally with each other (the dashed line), their formal work communications flow through the service manager.

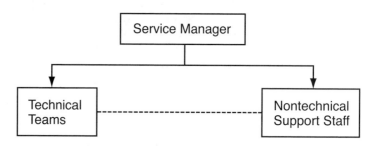

FIGURE 3-6 A formal working relationship.

Leadership Expectations

There are numerous leadership strategies for team leaders to consider. Service consultants should spend some time studying the different leadership styles.

In general, a team leader at a service facility has a number of major responsibilities. Several of these are:

- the facility's profit status
- the quality of the service being provided
- customer satisfaction
- team efficiency
- employee morale
- the reputation of the service facility

There are many other leadership responsibilities, but the point is that the monitoring and managing of these responsibilities is the function of the team leader. It is not a job that can be learned overnight.

In addition to these responsibilities, the leader has numerous other duties. For example, the team leader is required to set a good example with respect to personal appearance and conduct. Further, leaders cannot expect subordinates to behave in a manner that they do not personally practice. This includes the maintenance of composure, use of language, interactions with other people, being on time, taking one's job seriously, honesty, respect for others, and so on.

Another responsibility of the team leader is the **cross training** of team members. People should learn each other's jobs. Of course, this cannot be done for a job that requires a lot of education and training, but employees should be able to perform other job duties at the facility and cover for each other when someone is absent. For example, in a small independent facility, the service consultant should know how to order parts and prepare an estimate. If a staff member, such as the cashier, is not available, the service consultant, parts specialist, and lead technician should know how to receive a payment from a customer. In this case a set of detailed directions may be needed for the use of the credit card machine.

Another duty for the team leader is team building. There are many recommendations made by experts on team building and the team leader should become familiar with them. The purpose of team building is to have the team members know each other well enough (even if they don't like each other) to work together like a well-oiled machine, meaning with little or no friction.

In addition, proper and clear communication among team members is critical. A team leader should be focused on the ability of team members to communicate. Nothing can cause problems quicker than if people do not understand each other. Communication may include written and verbal exchanges. Handwriting is important when ROs are exchanged. For example, when a part is ordered for an automobile, the year and model of the automobile and the description of the part must be written clearly on the RO. If handwritten messages are not clear, the process may be dramatically delayed. Furthermore, some jobs may require knowledge of a foreign language because some customers and employees may not speak English. For example, in some states such as Florida, the ability to speak Spanish has become a job requirement at some service facilities.

Team leaders must always be concerned with weaknesses in the process and in team members. This requires them to be observant and to mentally review the activities at the end of the day after the facility closes. When weaknesses are noted, team leaders must correct them immediately. This should be done in a manner that corrects the problem and does not make it worse or create another.

Finally, a more successful leadership strategy is for the leader to treat team members in a respectful manner. Maintaining personal dignity and respecting the dignity of others have been found to be more successful than a top-down, rigid, and controlling style of leadership, especially if people are expected to work as a team with the members supporting each other and the leader.

The Use of Flat-Rate Objectives

When a facility uses **flat-rate objectives**, the team leader, who may be the service consultant or lead technician, must monitor them. A flat-rate objective is the number of flat-rate hours a technician hopes to earn by the end of the week. For example, at the end of the beginning of each quarter each technician meets with management to decide how many flat-rate hours he or she hopes to earn each week. The amount is based on the technician's past performance with input from the lead technician about the type of work the technician is capable of performing. In other words, a paper declaration is not enough since the flat-rate objective must be discussed and be based on a performance review.

Measurement of how close each technician is to his or her objective is provided to the team leader each morning by the service manager or consultant in a daily objective report. The report contains each technician's name as well as the flat-rate objective and the actual amount of time he or she had the previous day (see Figure 3-7). Figure 3-7 shows that technician 1 did not reach his or her target while technicians 2 and 3 either reached or exceeded their target. This flat-rate report should be used as a reference when the lead technician assigns work, assuming that the leader is attempting to help each technician reach his or her objective.

The lead technician examines the flat-rate objectives relative to the morning report of the flat-rate hours earned to determine what kind of work to give each technician. If the technician is running behind in the number of hours earned, for example, because of loss of time on a difficult

Technician		Day 1	Day 2	Day 3	Day 4	Day 5	TOTAL
1	Daily flat-rate total	9.1	3.8	11	13	10.1	47
	Daily flat-rate target	10	10	10	10	10	50
	Ahead or behind	−0.9	−6.2	1	3	0.1	**−3**
2	Daily flat-rate total	8	8	8	8	8	40
	Daily flat-rate target	8	8	8	8	8	40
	Ahead or behind	0	0	0	0	0	**0**
3	Daily flat-rate total	10	10.3	9.4	9.8	9.9	49.4
	Daily flat-rate target	9.5	9.5	9.5	9.5	9.5	47.5
	Ahead or behind	0.5	0.8	−0.1	0.3	0.4	**1.9**

FIGURE 3-7 A daily labor report for flat-rate objectives.

job, the lead technician may give the technician some easier work so he or she can make up for the time lost. When a technician is ahead of the flat-rate objective, he or she may be assigned work that is more difficult as long as the technician has the ability to do the job.

Summary

As the automobile evolves and contains more and more complex systems, the automobile service facility must hire personnel who can work on them. Employees with the expertise needed to work on these complex systems must work on them and not on other jobs that can be performed by a lower skilled/lower paid employee.

In addition, as customers pay more and more of their personal income to purchase these sophisticated automobiles, they expect efficient and professional service for their maintenance and repair. An automobile service facility must meet customer expectations to stay in business. If it does not, cannot, or will not, another facility will.

Today's automobile service facilities have had to change to meet the expectations and demands made on them. One of the methods to successfully meet the ever-changing expectations and demands has been to hire people with the variety of knowledge and skills needed to take care of customers, to work on the different automobile systems, to take care of the facility, and to manage the business operations. Next these employees must work together, or, as some people say, "they must be on the same page." To accomplish this, the employees are considered to be a team, regardless of whether there are two people or twenty people.

Two major missions for the service facility teams are to satisfy the customers and work together efficiently. If these missions are ignored, the service facility will not stay in business. In addition, to meet the expectations of customers and to be efficient, the number of team members and their relationships must fit the operations of the facility. Team arrangements depend on the size and type of the facility. For example, larger facilities require more team members with different working relationships than a smaller facility. Also an independent service facility has a different set of relationships than a facility in a retail store, which employs a store manager.

Understanding the differences in teams is important to the service consultant. In many cases, the service consultant is the sole team leader, while at other service facilities the leader responsibilities are shared with a lead technician. In addition, while the size and type of facility and the team arrangement require some of the same leadership responsibilities, there are differences to be recognized. For example, in some facilities the technicians are paid an hourly rate, at others the facility uses the flat-rate method, and some have daily flat-rate objectives that must be monitored.

Therefore, the team approach with a team leader should provide a service facility with a self-motivated group of professionals who work together for the benefit of the facility. If the facility is successful, the team members will be successful. Both can then share in the security and benefits of success.

Review Questions

Multiple Choice

1. _____ must be able to enter a facility with minimal effort and confusion.
   A. Customers
   B. Technicians
   C. Service consultants
   D. Managers
2. According to the textbook, the team's mission is to:
   A. work together to ensure customers are treated as important people and then satisfy them with the service and treatment desired
   B. work together to ensure the efficiency of the facility
   C. both A and B
   D. neither A nor B

3. A typical team has:
   A. a service consultant, technicians, and a manager
   B. a service consultant, a manager, and a parts specialist
   C. a service consultant, technicians, and a parts specialist
   D. a manager, a customer, and a parts specialist
4. At a service facility with a lead technician, the work is assigned to technicians by:
   A. the service consultant
   B. the lead technician
   C. the manager
   D. the customer
5. When technicians are paid by the job, it is called:
   A. the flat-rate system
   B. the hourly pay system
   C. the labor guide system
   D. the technician system
6. Technician A says that a flat-rate objective is the number of flat-rate hours a technician hopes to earn by the end of the week. Technician B says that he sets his own flat-rate objective and management does not have any input.
   Who is correct?
   A. A only
   B. B only
   C. Both A and B
   D. Neither A nor B

Application Problems

Service Consultant Math Exercises: Calculate the following diagnosis charges based on a labor rate of $80 per hour.
   A. Diagnose engine misfire (1.0 hour)
   B. Road test and confirm noise (0.3 hour)
   C. Diagnose a no-start, no-crank concern (0.6 hour)
   D. Check computer codes and perform simple circuit checks (1.3 hours)
   E. Perform simple electrical circuit checks (0.5 hour)

Calculate the ahead or behind totals for F through K

Technician		Day 1	Day 2	Day 3	Day 4	Day 5	TOTAL
1	Daily flat-rate total	9	8	11	7	10	45
	Daily flat-rate target	10	10	10	10	10	50
	Ahead or behind	F	G	H	I	J	K

Short Answer Questions

1. Explain why and how a team approach can offer superior customer service.
2. Present optional team formations and team member assignments in different types of service facility operations.
3. Identify the major responsibilities of the team leader.

CHAPTER 4

CHECKING VEHICLE AND CUSTOMER RECORDS

OBJECTIVES

Upon reading this chapter, you should be able to:

- *Explain why an automobile service history is important (Task A.1.9).*

- *Define repeat repairs/comebacks (Task D.6).*

- *Describe how an automobile repair history is recorded and stored (Tasks A.1.2, A.1.4, and A.1.10).*

- *Explain the importance of first-time, warranty, repeat repair, fleet, and regular customers at a service facility (Task A.1.13).*

Introduction

A person's history of treatment, whether it is with a family physician or an automobile service facility, is very important to him or her. As a result, people expect professionals, such as a medical doctor, to maintain records of all of their illnesses and examinations as well as the treatment and the results of the treatment.

Likewise when people are regular customers at an automobile service facility, they expect the service consultant to have records of past services performed on their vehicle. This is expected even if the service consultant is a new employee at the facility and has never met the customers. Therefore, maintaining a history of services performed on customer automobiles and informing them about it is important to a service facility that wishes to present a professional image.

In addition, knowing the history of automobiles serviced at a facility is important to track its earnings and losses. For example, at Renrag Auto Repair a customer called and reported that her automobile's cooling system was overheating. She was quite distressed because she had brought her automobile into the shop for the same problem four weeks earlier. The service consultant pulled her name and vehicle up on the computer, verified that the vehicle had been in for a cooling system repair, and asked her to bring her vehicle in right away.

Upon checking the customer's record in the computer, the service consultant learned that the water pump had been leaking and was replaced. Next the service consultant examined the comments on the invoice and the original comments on the technician's worksheet (also called the repair order or hardcopy). One recommendation recorded on the invoice by the consultant was to replace the thermostat. This was recommended because of the technician's experience that this part had a high failure rate and when stuck closed would cause overheating. In addition, the cost to replace the thermostat when the water pump was replaced was low compared to replacing it later. The written comments on the invoice indicated that the customer was advised of the recommendation but declined to have the thermostat replaced even though it might prevent a future breakdown.

When the customer brought her automobile into the service facility, the technician diagnosed the cooling system and found that the water pump was working properly. Further diagnosis found that the thermostat was broken and stuck closed. When the technician's diagnosis was presented to the customer, she was upset. The consultant then showed her the comments on the invoice at the time the water pump was replaced. She recalled the conversation and her response and knew that she had made the wrong decision. Although she was not happy to pay for the repair because it would have cost her less money a few weeks earlier, she left with a positive attitude about the service facility, the competence of the technician who worked on her car, and the detailed records kept on her car.

There are several points to be recognized in the preceding example. To start, the service consultant had two records to check: the invoice stored in the computer database and the technician's hardcopy notes in the facility's filing cabinet. Without this history, Renrag would have had a customer relations problem and possibly would have had to pay for all or a percentage of the repair charges.

Regular and First-Time Customers

Service consultants work with different types of customers each day. The most important are the regular customers and people who may become regular customers. Regular customers have all of the work performed on their automobile in the same facility. They represent a steady income. This means that over time, they spend a considerable amount of money at a facility, and, as a result, influence its sales volume and profit.

Regular customers, however, are hard to recruit. Often a new service facility has to spend a lot of money for several years to establish a positive reputation in order to build a strong regular customer base. If regular customers are not treated properly, however, a service facility can lose them in a short period of time. Service consultants must also realize that regular customers occasionally "shop around" for better service and lower prices, especially if they think the service facility is not taking care of them. So service consultants must not take customers for granted or they will likely lose them to the competition.

Therefore, one of the major jobs of service consultants is to recruit and retain regular customers. This requires greeting regular customers in a personal manner and identifying potential regular customers when they enter the facility.

For example, assume that a regular customer, Mrs. McCord, likes to have the tires rotated on her automobile every other oil change. When Mrs. McCord arrives for an oil change, the service consultant should check her automobile's service history and determine if it should have the tires rotated. The service consultant should inform Mrs. McCord when her tires were last rotated and then ask if she wishes to have this rotation service. This shows Mrs. McCord that a record is being maintained on her automobile by the service facility and of the service consultant's interest in her preferences. She will most likely be pleased with the special treatment. The identification and recommendation of maintenance needs can help to retain a regular customer and make a new customer a regular one (Task A.1.10).

The First-Time Customer

When new customers are identified by the service consultant, the objective is to convince them that the service facility can help them take care of their automobile. One approach to demonstrating this willingness to provide assistance is for the service consultant to place a flag on the repair order to let the technician know that this is the customer's first visit to the facility.

During the customer's first visit to a facility, the technician should conduct a detailed inspection of the general condition of the automobile to identify anything that is wrong with it. This inspection is similar to that conducted on all automobiles being serviced, except that it is more thorough. At Renrag Auto Repair, the technicians filled out a form used for a general inspection of all automobiles (see Figure 4-1) and a more detailed form for new customers (see Figure 4-2).

When new customers are advised of the inspection and of anything found that needs attention, the service consultant should be able to offer them a complementary service such as an oil change or tire rotation. This offer shows that the service facility is interested in them and wants to demonstrate to the customer that they can help them maintain their automobile. Then the service consultant should record the information collected on the automobile to create the historical database on the automobile. New customers, as well as the regular customers, should be given a copy of the inspection form.

Automobile History

An automobile service history is a record of the diagnosis, repairs, and maintenance performed on a customer's automobile by the service facility. The records presenting the history are the estimates, repair orders, and invoices that are usually recorded in a computer database. Just as important, though, are the notes of the technician and service consultant on any hardcopies that should be kept on file. The old rule to remember is that the three most important duties of a manager are to: *(1) document, (2) document, and (3) document.*

The sets of estimates, repair orders, and invoices as well as the written documents on customers' automobiles are kept in the **shop management system**. The shop management system consists of the procedures, documents, files, and computer used to prepare, store, and retrieve customer information. Service consultants must have a thorough knowledge of the shop management system used by their facility. For example, service consultants must be able to file and retrieve hardcopies with handwritten notes and customer approvals, which should be stored in alphabetical order in the facility's filing cabinets. In addition, they must work with the computer program that stores the estimates, repair orders, and invoices. Furthermore, they must know how to back up the records in the computer database on an alternative storage device, such as a disk or tape, at the close of each business day. If the computer "crashes" and the facility loses all of the information on the hard drive, the work and cost required to re-create the database is extensive!!

Customer Name _____

Date_____

Vehicle Year _____ Make _____ Model _____

Mileage _____ Engine_____

State Inspection Sticker Date _____

General Exterior Inspection_____

_____

1. Exterior/Interior Lights_____

2. Windshield Wipers/Blades/Washers _____

3. Belts: OK   Cracked/Worn _____

4. Antifreeze:

      Acid Strip Test for pH: OK   Recommend Flush

      Temperature Strip Test: _____ degrees

      Color: OK   Recommend Flush

5. Radiator Hoses: OK   Soft/Cut _____

6. Battery: Date of Purchase _____

7. Transmission Fluid: OK   Topped Off   Burned Odor

8. Air Filter: OK   Recommend Change

9. Fluid Levels: OK   Low

10. Exhaust: OK   Leaks Noted

11. Struts/Shocks/Springs: OK   Leaks/Damages

12. Tires: Pressures at _____ lb

LF:	OK	Feathered	Worn to Wear Strips
RF:	OK	Feathered	Worn to Wear Strips
LR:	OK	Feathered	Worn to Wear Strips
RR:	OK	Feathered	Worn to Wear Strips

13. CV Boots: OK   Worn or Torn

14. Steering Boots: OK   Worn or Torn

15. Oil/Transmission/Power Steering/Brake Fluid Leaks:

    None Noted   Leaks Noted at _____

16. Rear Differential: OK   Leaks Noted

17. Oil: # of Quarts _____

Recommendations: _____

_____

_____

FIGURE 4-1 A general maintenance inspection form.

Customer Name _____ Date_____

Vehicle Year _____ Make _____ Model _____

Mileage _____ Engine _____ PA Inspection Date _____

General exterior inspection (checkmark = OK; N = not OK):
_____ 1. Lights & lenses   _____ 2. Turn signals
_____ 3. Brake lights   _____ 4. Headlights (hi/low)
_____ 5. Body (general)   _____ 6. Glass
_____ 7. Wipers   _____ 8. Ext. mirrors
_____ 9. Bumpers   _____10. Doors
_____11. Gas cap   _____12. Hood
_____13. Trunk   _____14. Shocks/struts
_____15. Hubcaps/wheels

Under the hood (checkmark = OK; N = not OK):
_____ 1. WW washer   _____ 2. Belts
_____ 3. Hoses   _____ 4. Antifreeze Acid Strip Test for pH
_____ 5. Antifreeze Temperature Strip Test _____ degrees
_____ 6. Antifreeze color   _____ 7. Antifreeze level
_____ 8. Air filter   _____ 9. Hood latch
_____10. Master cyl.   _____11. Transmission fluid
_____12. Firewall holes

Electrical (checkmark = OK; N = not OK):
_____ 1. Battery terminals   _____ 2. Battery hold down
_____ 3. Starter draw test   _____ 4. Battery load test
_____ 5. Charging system test   _____ 6. Ignition system
_____ 7. Computer code check

Underside (checkmark = OK; N = not OK):
_____ 1. Shocks/struts   _____ 2. Springs
_____ 3. Ball joints   _____ 4. Steering
_____ 5. Boots-CV/steering   _____ 6. Floor/frame
_____ 7. Powertrain mounts   _____ 8. Fuel system
_____ 9. Exhaust
_____10. Tires: RF ____ , LF ____ , RR ____ , LR ____
_____11. Brakes: RF ____ , LF ____ , RR ____ , LR ____
_____12. Rotors/drums RF ____ , LF ____ , RR ____ , LR ____
_____13. Cylinders & calipers   _____14. Hoses & lines
_____15. Oil leaks—powertrain   _____16. Coolant leaks
_____17. Oil leaks—power steering

FIGURE 4-2 A special maintenance inspection form.

Inside (checkmark = OK; N = not OK):

_____ 1. Steering	_____ 2. Horn
_____ 3. Mirror	_____ 4. Brake pedal
_____ 5. Emergency brake	_____ 6. Seats
_____ 7. Safety belts	_____ 8. Accelerator
_____ 9. Hood latch	_____ 10. High beam indicator
_____ 11. Turn signal indicator	_____ 12. Inside lights
_____ 13. Power acc. check	_____ 14. Heater check
_____ 15. A/C check	

Recommendations: _____

_____

_____

FIGURE 4-2 A special maintenance inspection form. [continued]

Computerized Automobile Records

Customer service histories are often stored in a computer database. In some cases, these records are referred to as service records or service files. To create a record, the service consultant has to enter each customer's information into the computer database. These entries consist of the customer's name, address, phone number, and automobile information. Often this information must be entered into the database before an estimate or repair order can be generated so that work can be performed on a customer's automobile.

Creating a Customer File

The service consultant begins a file for each customer in the database when the customer initially contacts the service facility. This often occurs when a customer calls to make an appointment. Therefore, the computer and phone have to be located at the service consultant's workstation (see Figure 4-3).

The creation of a customer's file is usually completed when he or she arrives and has been properly greeted by the service consultant (Task A.1.6). At this time any information not obtained in the phone conversation is entered into the computer database file. When the customer's information has been entered into the database, many computer programs

FIGURE 4-3 A service consultant's workstation with a phone and a computer.

will permit the service consultant to schedule an appointment, produce an estimate, generate a repair order, and print a final invoice.

After the customer's information has been entered in the database, the service consultant can very quickly retrieve the automobile history. To retrieve this information, the service consultant may simply type in one of the following:

• the customer's last name and automobile
• the vehicle identification number
• the vehicle's license plate number

Upon calling up a customer's automobile record to the computer screen, the service consultant may review its service history. The service consultant must be cautioned, however, to be sure that the automobile is the correct one. For example, at Renrag, a customer purchased a new automobile that was the same make, model, and color as the old one, and another man and his wife had the same year, make, and model of automobile. In these cases, when the service was recorded and the mileage of the automobile was entered into the computer, the service consultant became aware of a problem.

Some computer programs allow for a search on all automobiles in the database that are of a specific year, make, and model. A list of all of these automobiles is then shown on the computer screen along with the owners' names. A service consultant may then be able to identify a customer's last name in order to open the correct file. This is handy if the service consultant has obtained information on an automobile but has misplaced or cannot recall the name of the owner.

In addition, a list of automobiles by model may be particularly useful when buying supplies for inventory. For example, some oil filters only fit certain automobile engines. If the facility's database shows that no customers have an automobile with a specific type of engine, it does not have to order any of the filters for that engine for the shop inventory. Of course, inventory orders of filters for engines in more popular models can be based on the number of automobiles in the database.

Preparing Estimates, Repair Orders, and Invoices

To comprehend how the computer database is used to create a document for the service consultant and technicians, an overview of the process is shown in Figure 4-4. The process begins when the customer enters the service facility and the service consultant must create a database file.

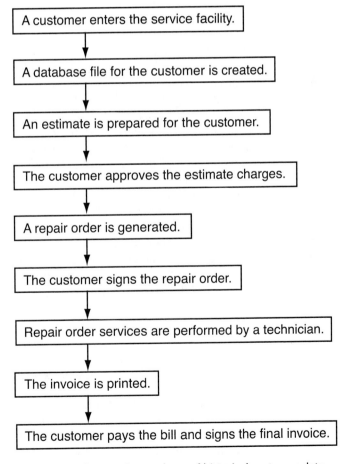

FIGURE 4-4 The creation and use of historical customer data.

After the necessary information is entered into the computer database, an estimate for the cost of a repair or maintenance is prepared, as discussed in the previous chapters and shown in Figure 4-4. Most state consumer protection laws require the estimate to include the parts and labor costs plus any additional fees, such as towing and tax. Most computer programs can make the necessary calculations for the cost of the parts, labor, additional supplies, other charges, and taxes. The computer then prints the estimate for the customer's review. Typically, an estimate by a service facility is valid for 90 days, although it may be more or less depending on the parts needed and state law.

After a customer approves of the work to be performed on his or her automobile, as shown in Figure 4-4, the service consultant prepares a repair order. The computer prints the repair order that lists all of the work to be performed, the parts to be used, any other charges for the service, the tax to be paid, and the total cost. The printed repair order must be signed by the customer, and the service consultant should indicate what has and has not been approved and the date and the time of the approval, and then also sign the order. If the customer is not at the facility and the estimate is approved over the phone, the service consultant must write the name of the person (typically the owner of the automobile) approving the service on the repair order, what was approved and not approved, and the date and the time of the approval, and then sign it. In both cases, the approved repair order becomes a binding contract between the customer and the service facility.

After the work is completed, a third document is prepared by the computer. This document is an invoice (see Figure 4-4) that lists all of the work performed, the parts used, any other charges for the service, the tax to be paid, and the total amount to be paid by the customer. Depending on state law, the final invoice may need to be equal to or less than the estimate. If the invoice amount is going to be more than the amount approved on the work order, the customer must have agreed to the additional charges in writing or in a phone conversation that is documented by the service consultant.

A good customer relations practice is for the final invoice for a repair to be less than what the customer expects to pay. Experienced service consultants know that when the final invoice charge is more than what a customer expects to pay, he or she is likely to become upset and dissatisfied with the services performed regardless of the quality of the work.

Dealerships and Warranties

Automobile manufacturers and dealers prefer to have the people who buy new cars from them to continue to purchase their new automobiles there, so customer satisfaction with the repairs and services is extremely important. Therefore, dealership service departments perform a large number of manufacturer warranty repairs and free maintenance work, such as oil changes, for customers who buy new automobiles from them. To ensure that their customers are satisfied, the automobile manufacturers survey their new car customers about the services they receive. If the survey results are not positive, the sale of new automobiles can be affected. The consequences of negative results of a survey at a dealership service department can range from the restructuring and retraining of personnel in the service department to disciplinary actions, such as the termination of an employee.

Another reason special attention must be given to warranty customers is because state laws, which are referred to as lemon laws and discussed further in Chapter 5, set limits on the services needed by a buyer. Specifically, these laws set a maximum number of visits a person who purchases a new automobile has to make to the dealership service department for the same repair service and/or the number of days a customer's automobile can be out of service before the dealer or manufacturer must buy it back. The service consultant must help the dealer and automobile manufacturer avoid any costly and troublesome buybacks of new automobiles by closely monitoring the number of visits for the same repair and the number of days a customer's automobile is out of service.

Comebacks

Repeat repairs are called **comebacks** or **second attempts**, meaning a customer had a repair performed at the service facility and must return because the same repair must be made again on the automobile. The reason for the return is typically unknown, and just as warranty repair customers are extremely important to a service consultant, so too are repeat repair customers. When a customer returns an automobile because the repair is thought to be unsatisfactory, the service consultant must express an interest in examining and correcting the problem. The customer should be advised that after a thorough diagnosis by a technician, a plan of action will be prepared. The service consultant should never propose that a charge for the repair will or will not be made until the automobile is examined.

The key to handling comebacks is for the service consultant to collect the information about past services conducted on the automobile. When the past records of an automobile are retrieved, the consultant must identify when the service was performed and the number of miles that the automobile has been driven since the service was performed. Date and mileage are important because they usually determine whether the customer will be charged for the repair. Next, the service consultant must review the notes and comments about the repair from the hardcopies in the file.

There is no standard number of miles traveled or days that have lapsed after the repair was made to determine whether a customer will be charged for a repeat repair. Often the determining factor is the service facility's policy, an automobile manufacturer warranty policy (discussed further in Chapter 5), a state consumer protection law, or the parts warranty. Service consultants must be aware that the attorney general in their state must enforce the consumer protection laws assuring customers that they are protected from unfair treatment when their automobile is serviced.

Therefore, the consumer protection laws have guidelines that set forth the conditions related to automobile repairs. To understand the law and the position of the Office of the Attorney General, service consultants must obtain a copy of the consumer protection laws from it or from its Web site. Upon reading the consumer protection laws, service consultants should work with their managers and facility owners to ensure that their service facility's policies, procedures, advertising, and promotional campaigns do not violate the law. For example, if a service facility states that a customer will receive a "money back guarantee if not satisfied," then the state consumer law will likely define what that means. In Pennsylvania, when the customers are not satisfied their money must be returned within five days.

Parts Warranty

When a comeback is caused by a defective part, many parts suppliers have a warranty that covers the cost of the new part. In some cases, this warranty may cover the labor cost to replace the part. The parts warranty should state how long and under what conditions a repair will be made free of charge. Therefore, service consultants must know the terms of all of the warranties issued by their service facility.

For example, assume a parts supplier offers a one-year warranty on all parts and labor on remanufactured alternators. This means that neither the service facility nor the customer should have to pay anything if the alternator needs to be replaced within the warranty time period. However, in some cases a warranty may cover only a percentage of the costs after the car has been driven a specified number of miles or days. In this case, if the service facility has to charge the customer for the amount not covered by the warranty, then he or she must be advised of the difference. If not, the service facility must pay the difference.

Working with Fleets

Fleet customers own a few to several dozen automobiles that need regular service. A business, such as a construction company, or a government agency, such as a police department, may own a fleet of automobiles serviced by one facility. The fleet is similar to repeat customers because it provides a considerable percentage of the sales made by a service facility. When providing fleet maintenance services, it is common to have prearranged agreements about when the work is to be performed. For instance, in many cases certain maintenance tasks are to be automatically performed when an automobile is at or beyond a certain mileage. If the facility also repairs the fleet automobiles, the usual expectation is that their vehicles will receive priority service.

Unlike other customer automobiles, fleet automobiles that need work are either dropped off by a fleet employee, who may or may not be the regular driver of the automobile, or picked up by an employee of the service facility. As a result, when the maintenance or repair of a fleet automobile is needed, the service consultant may not have the opportunity to talk to the driver of the automobile to obtain information about the problem or the performance of the vehicle. Rather, this information may come from the fleet manager or through some other means, such as a note left in the automobile.

Although fleet customers may be desired by a service facility, care must be taken because many fleets require service discounts and special treatment such as "on-demand" and evening service, and often take 30 days to pay their invoices. For larger service facilities that have extended hours and extra technician flat-rate hours to sell, this may not be a problem. However, for smaller service facilities, fleets can cause cash flow problems and operational headaches.

At Renrag Auto Repair, for example, one of the fleet service customers that had over fifteen vehicles received a discount on all services and often had to have work conducted in the evening or on weekends. The schedule for their maintenance is shown in Figure 4-5. To meet this service

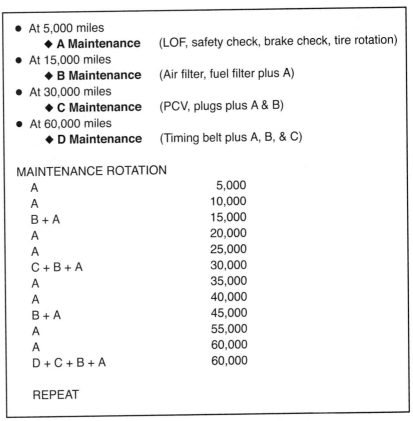

- At 5,000 miles
  - ◆ **A Maintenance**   (LOF, safety check, brake check, tire rotation)
- At 15,000 miles
  - ◆ **B Maintenance**   (Air filter, fuel filter plus A)
- At 30,000 miles
  - ◆ **C Maintenance**   (PCV, plugs plus A & B)
- At 60,000 miles
  - ◆ **D Maintenance**   (Timing belt plus A, B, & C)

MAINTENANCE ROTATION

A	5,000
A	10,000
B + A	15,000
A	20,000
A	25,000
C + B + A	30,000
A	35,000
A	40,000
B + A	45,000
A	55,000
A	60,000
D + C + B + A	60,000

REPEAT

FIGURE 4-5 A fleet maintenance schedule.

schedule, employees had to be paid overtime, assuming they would be available to work, and additional part-time employees had to be hired to shuttle the vehicles to and from the location where the fleet was parked. As a result, even though the fleet provided a lot of business for the facility, the discount demands, overtime pay, and additional costs became too expensive and a loss on the service was incurred.

In addition, when fleet customers have to have 30 days or more credit, problems can be created depending on the size of the facility and the fleet, and the amount of money the facility must spend before getting paid. For example, most employees are paid weekly and other charges, such as the parts bills, are paid at the end of the month, although some parts must be paid when they are delivered. This means the facility must be able to finance the maintenance and repair of the fleet vehicles. This may cause a cash flow problem because the facility may have to wait up to 60 days to receive a payment from the fleet for a service. For example, if the fleet has a service at the beginning of the month, the invoice will not be sent until the end of the month (30 days) and then the fleet may take another 30 days to pay. As a result, when making a contract with fleet managers, service consultants should consult with the facility owner or manager before an agreement is reached.

Summary

The ability of a service consultant to check a vehicle and customer service record depends on the records maintained by the service facility. In turn, the usefulness of the historical documents depends on the facility's shop management system. There is a wide range of systems available. Some are quite basic and limited while others are complex and elaborate. The selection of a system is typically dictated by the facility's size, such as the volume of business and availability of personnel to use the system's program.

Regardless of the size of a facility or the features of the shop management system, the service consultant must be able to use it. Therefore, service consultants must be aware of the different types of programs that are available for service facility systems. Further, they must also keep up-to-date as new programs come on the market or as existing programs are modified. A problem may arise when a facility's shop management system is out of date and no longer competitive. Likewise service consultants must stay up-to-date or they will not be competitive. Worse yet, service consultants who cannot or will not learn to use the latest technology will eventually be replaced with those who can or will use it.

Review Questions

Multiple Choice

1. Service consultant A says that when writing up a second attempt (comeback)/ warranty ticket, it is necessary to review previous repair orders. Service consultant B says that when writing up a second attempt (comeback)/warranty ticket it is necessary to ask the customer to state the symptoms he or she is experiencing.
   Who is correct?
   A. A only
   B. B only
   C. Both A and B
   D. Neither A nor B

2. A vehicle in the shop for an oil change shows approximately 59,000 miles on the odometer. What should the service consultant do?
   A. Suggest an appointment for a 60,000-mile maintenance service.
   B. Offer the customer a discount to perform the 60,000-mile service today.
   C. Advise that the 60,000-mile service is covered by the manufacturer's warranty.
   D. Provide a ballpark estimate for a 60,000-mile maintenance service.

3. Each of these represents an example of customer information that might be included on a repair order EXCEPT:
   A. an e-mail address
   B. a preferred payment method
   C. a cell phone number
   D. the service consultant's name

4. The technician notes on the repair order that the fuel filter appears to be the original one on a vehicle with almost 60,000 miles on it. The item calls for replacement at 30,000 miles. Which of these should the service consultant do?
   A. Estimate a maintenance tune-up including the fuel filter.
   B. Ask the customer when and if it was replaced.
   C. Tell the customer that it has not been replaced in 60,000 miles.
   D. Leave the item for the 60,000-mile service.

5. A customer calls and states that his or her vehicle has a problem that has had several repair attempts. Which of the following should the service consultant do?
   A. Determine if the dealership/shop has ever worked on the vehicle.
   B. Offer to take the vehicle in immediately.
   C. Ask the customer to provide previous work orders.
   D. Explain that sometimes a problem can take several attempts to resolve.

Short Answer Questions

1. Explain why automobile service history is important to both the service facility and the customer.
2. What are repeat repairs/comebacks? (Task D.6).
3. How is an automobile repair history recorded and stored?
4. How are first-time, warranty, repeat repair, fleet, and regular customers different?

CHAPTER 5

WORKING WITH WARRANTIES, SERVICE CONTRACTS, SERVICE BULLETINS, AND CAMPAIGNS/RECALLS

OBJECTIVES

Upon reading this chapter, you should be able to:

- *Locate and use reference information for warranties, maintenance contracts, and campaign recalls (Task B.6.2).*

- *Define warranty policies and procedures/parameters (Task B.6.1).*

- *Verify the applicability of warranties, maintenance contracts, technical service bulletins, and campaigns/recalls (Task B.6.4).*

- *Explain warranty, maintenance contract, technical service bulletin, and campaign/recall procedures to customers (Task B.6.3).*

Introduction

Not all customers at an automobile service facility personally pay for their maintenance or repair; sometimes, payment is covered by a warranty, a maintenance contract, or a recall/campaign repair order. When working with these customers, service consultants are required to know (1) what each type of contract represents; (2) why manufacturers use them; and (3) how to work with them.

Although warranties, maintenance contracts, and recall/ campaign repair orders may not make up a large volume of the work conducted at an independent auto service facility, dealership service departments do a considerable amount of this type of work for automobile manufacturers. In both facilities, therefore, the service consultant must know how to take care of the customers' contracts or they could lose regular customers and the reputation of the facility could be harmed.

Specifically, service consultants must be knowledgeable about the various types of contracts because payments depend on whether the automobile is properly serviced according to the guidelines stated in the agreement. Therefore, the purpose of this chapter is to define, describe the purpose of, and explain the procedures related to the use of warranties, maintenance contracts, and recall/campaign repair orders.

Definitions and References for Warranties, Maintenance Contracts, and Campaigns/Recalls

A **warranty** contract functions much like an insurance policy. A claim is made to a warranty company that will pay for the vehicle repair if the terms outlined within the warranty contract are met. In simple terms, customers buy a warranty contract because they are betting that a covered component may break before the warranty coverage period expires, and the warranty company is betting that the covered component will not break.

A **maintenance contract** (also referred to as a service contract) is often bought by (or in some cases given to) a customer when an automobile is purchased. The contract pays for maintenance services the automobile needs for a specified coverage period. This is like prepaying for the maintenances an automobile will need in the future. When used regularly at the mileage intervals specified in the contract, customers often save money and are assured that their automobiles stay in top running condition.

Finally, **recall/campaign** repairs are often the result of a **National Highway Traffic Safety Administration (NHTSA)** or an **Insurance Institute for Highway Safety (IIHS)** investigation that found a problem with a certain year, make, and model of an automobile. The manufacturer is required under penalty of law to repair the problem for the consumer free of charge.

Eligibility

To determine whether an automobile is eligible for warranty or maintenance contract work, the service consultant should ask customers if they own one. If they do, the service consultant should ask for it.

In some cases, a phone number for contract assistance is available, particularly if the contract was purchased from a source other than a new automobile dealership. When the contract is purchased from a new automobile dealership, it is common for its service consultant to be able to access the information from the automobile manufacturer's database. This is because most new automobile dealerships have a computer that is linked to the manufacturer. Unfortunately, a service consultant at an independent service facility may not have access to this same information; in this case the customer has to contact the dealership or manufacturer's hotline for specific details about the contract.

Recall and campaign information for a specific automobile can be obtained through the manufacturer's database by dealership service consultants and technicians. Automobile owners can sometimes obtain information about their automobile by calling the manufacturer's customer hotline. In some cases, owners who are in the manufacturer's database will be notified by mail about a recall on their automobile. Service consultants at independent service facilities can often obtain recall information from a:

- computerized database (Mitchell-On-Demand and Snap-on's Shop Key system, among others).
- publication (Automotive News or the Automotive Service Excellence publication *Tech News*).
- Web site (the NHTSA at http://www.NHTSA.dot.gov or the Insurance Institute Web site at http://www.highwaysafety.org).

In addition to recalls and campaigns, automobile manufacturers release repairs for other concerns. Instructions about how to repair these concerns are described in a manufacturer's publication called a **technical service bulletin (TSB)**. Manufacturers release TSBs regularly for dealership service department technicians. Many computer data systems purchased by independent service facilities can provide TSB information to technicians at independent service facilities as well.

Warranty Policies and Procedures

There are four types of automobile warranties that service facilities and the service consultant should recognize and administer. The categories are:

- new automobile warranty contract
- bumper-to-bumper warranties
- extended warranty contracts
- emission warranties

New Automobile Warranty Contracts

A **new automobile manufacturer warranty** contract typically applies to new automobiles sold to the consumer. The warranty provides for the repair of the automobile "free of charge" (in some cases there is a deductible discussed later) provided the automobile is within a predetermined time frame and under a predetermined mileage. This time period is called the **warranty coverage period**. The warranty coverage period may be 12 months from the time of purchase, but it can be as long as several years and 100,000 miles (typically with certain conditions) for some manufacturers.

Customers with automobiles covered by a manufacturer's warranty must take their vehicles to a service facility that is authorized by the manufacturer. This is typically the dealership service department where they purchased their automobile. This does not mean, however, that they must use the same dealership where they bought their automobile.

The predetermined mileage set forth in a new car warranty, for example, may be 12,000 miles or higher depending on the manufacturer and the state's laws. Sometimes a customer buys a "factory authorized" pre-owned automobile from a new automobile dealer and the balance of the predetermined mileage may be passed on to the new owner.

When a customer's automobile is covered by a new automobile warranty, the diagnostic charges, repair charges, and parts are provided free of charge. This is assuming the automobile is within the warranty coverage period and the repair is not a result of customer abuse or an accident. For example, one customer drove his automobile at high speed through a flooded street and got water into the engine through the air intake. The connecting rods inside the engine were bent because they could not compress the water. The owner's misuse of his automobile caused the damage. Therefore, the manufacturer did not cover the cost of the repairs, even though the automobile was within the warranty coverage period and the warranty covered the automobile's powertrain.

For a warranty to be valid, the automobile must be under both the mileage and the time interval limits. For example, if an automobile owner with a 12-month, 12,000-mile warranty has a covered repair but the automobile has over 12,000 miles on it or he or she has owned it for more than 12 months, the warranty will not pay for the repair. However, in practice if an automobile is just slightly outside the warranty coverage period limits, many manufacturers often cover some, if not all, of the repair because they know that each customer represents many thousands of dollars in repeat business.

Therefore, when a situation dictates, manufacturer warranties often have procedures a service consultant can follow to request a warranty coverage period waiver for "special circumstances." In some cases the manufacturer's procedure may require the customer to pay for the repair. Then the claim will either be given to a manufacturer's representative when he or she visits the dealership or be sent to the manufacturer's warranty audit department for review. If approved, the customer is reimbursed for the payment of the repair.

Bumper-to-Bumper Warranty

When a manufacturer's warranty covers all of the systems and parts on the automobile, it is referred to as a **"bumper-to-bumper warranty."** This means all of the automobile components are covered for a specific period, such as the first 12 months or 12,000 miles. In recent years some manufacturers have sold automobiles with limited bumper-to-bumper warranties that exclude certain components after a specified mileage or time interval has passed. Therefore, some contracts may cover only selected systems and parts for an additional period, such as the last 12 months or 12,000 miles in a 24-month or 24,000-mile contract.

To be more competitive in the market and to sell more automobiles, some new automobile manufacturers provide very long warranty coverage periods on selected parts. For example, a manufacturer may offer a powertrain warranty that covers engine and transmission parts for 7 years or 70,000 miles. Note that the powertrain warranty is included in the first 12 months and/or 12,000 miles of the bumper-to-bumper warranty.

In some states, the law may set the minimum number of months and miles a manufacturer must provide a warranty on all new automobiles, for example 12 months or 12,000 miles. Therefore, a 7-year or 70,000-mile powertrain warranty that covers certain engine and drivetrain (transmission, differential, and axle) failures, is like receiving an added 6 years or 58,000 miles past the initial warranty period. If, during this extended warranty period, repairs to the powertrain are needed, the warranty will pay for all, if not most, of the cost. However, if repairs to the brake system are needed in the extended time period, the warranty will not cover any of the costs.

Extended Warranty

In some cases, customers may elect to extend their warranty coverage beyond what is provided by the manufacturer (for new automobiles) or dealer (for pre-owned automobiles with a "used car" warranty). To do this, a customer buys what is known as an extended warranty. When purchased, the extended warranty provides coverage for certain systems and parts for a period that begins at the time the customer purchases the automobile.

For example, assume that Colin Williamson bought a 36-month or 36,000-mile extended warranty at the time he bought his 2-year-old pre-owned Jaguar with 26,100 miles on it from the Jaguar dealer. As a result, his Jaguar would be covered under the extended warranty for the next 36 months and up to 62,100 miles. The price Mr. Williamson would pay for the extended warranty is determined by the length of the coverage period and the system components covered. The longer the coverage period and the more components/ systems covered, the higher the price.

To help control warranty costs, the extended warranty contract will often stipulate where service can be performed. For example, when an extended warranty is purchased from a new automobile dealer, the contract will often require that the warranty repairs be performed by an authorized service facility. In this case, the authorized service facility would be limited to new automobile dealership service departments for the make of automobile the customer owns.

In addition to service facility limitations, extended warranties often have a deductible that the customer must pay toward the warranty repair. The deductible could be as little as $25 but typically is between $50 and $100. Therefore, if Edgar Farmer bought an extended warranty with a $75 deductible and his automobile needed to have a $500 repair performed that was covered by the warranty, he would pay $75 and the warranty company would pay $425.

A contract may also stipulate that the customer pay for noncovered service items, such as maintenance, certain repairs, and possibly the cost of the diagnosis. For example, assume that Pat McCormick owns an automobile with an extended warranty with a $25 deductible. Pat's engine was within the extended warranty coverage period when it developed a strange noise upon acceleration. The diagnosis of the problem cost $150 and was not covered by the warranty, while the repair cost of $600 was covered. In this case, Pat's extended warranty would cover the repair but not the diagnosis, and the deductible of $25 would be applied to the repair and not the $150 diagnosis. Pat would therefore owe the service facility $175 for the diagnosis and the deductible. The warranty company would then pay the service facility $575 for the repair.

The warranty contract does not cover the cost of a repair if customer neglect is the cause of the problem. For example, if the technician found that Pat's engine noise problem was due to engine sludge caused by infrequent oil changes, Pat would need maintenance records to show that he changed the oil within the time frame recommended by the manufacturer of the automobile. If he does not show that he maintained the automobile as recommended by the automobile manufacturer, the warranty company may refuse to pay the repair.

In addition, warranty companies will not pay for a repair if components not covered by the warranty contract are defective. For example, if Pat's warranty contract states that it does not cover idler pulley bearings that are bolted to the engine and the technician finds that the idler pulley bearing is the cause of the noise, Pat would have to pay for the entire repair.

Furthermore, warranties will not pay for repairs if the problem is caused by system modification. For example, if Pat changed the engine's stock camshaft to a high-performance camshaft and it is determined to be the cause of the engine noise, then Pat's warranty may not cover the cost of the repair.

Warranty contracts change slightly when customers buy an extended warranty from a used automobile dealer or from a company on the Internet. Although the warranty coverage period and deductible function like the extended warranties sold by new automobile dealers, the warranty often allows the customer to have the repairs done at a service facility he or she selects. At the same time, however, the warranty may set a maximum amount of money it will pay for a system's repair or a maximum labor rate it will pay per hour. When this occurs the customer must pay for any charges that are beyond the contract's maximum limits. For example, if Pat's repair was based on 10 hours at $60 per hour ($600) and the warranty company's maximum labor rate was $50 per hour ($500), Pat would have to pay the $100 difference plus any deductible.

Emission Warranty

Another type of warranty is the **emission warranty** mandated by the federal government. Emission warranties require new automobile manufacturers to repair problems associated with an automobile's emissions components for a stated period of time.

In practice, emission components are covered under the bumper-to-bumper warranty for the coverage period. After the bumper-to-bumper warranty expires, the key emission components, such as the catalytic converter and powertrain control module, are covered for a time period directed by federal law, such as 10 years and 100,000 miles. Customers can obtain emission warranty service when they go to a new automobile dealership service department for the make of the automobile they own. Emission warranty repair is provided free of charge to the customer when the automobile is within the warranty coverage period and the broken components meet the coverage requirements.

Verification of Warranty, Maintenance Contract, Technical Service Bulletin, and Campaign/Recalls

When a customer's automobile is covered by a warranty, the following information becomes important for the service consultant to verify before service is provided.

- Is today's date within the warranty's specified time frame limits?
- Is the automobile's current mileage less than the warranty's maximum mileage qualifier?
- Does the contract's VIN match the number on the automobile being repaired?
- Is the name on the contract the same as the customer's name?
- Is the system to be repaired covered by the warranty?
- If so, was abuse, modification by the customer, or neglect by the customer the cause of the failure?
- Does the contract specify a deductible that the customer must pay?
- Are diagnostic charges covered by the warranty?

If this information is verified to pertain to the warranty, the service consultant can allow the service facility to perform the repairs under the warranty guidelines.

When a customer's automobile is covered by a maintenance service contract, the following information becomes important for the service consultant to verify.

- Is today's date within the contract's specified time frame limits?
- If so, has the amount of time that has passed since the last maintenance been long enough?
- Is the automobile's current mileage less than the contract's maximum mileage qualifier?
- If so, has the number of miles since the last maintenance been long enough?
- Does the contract's VIN match the number on the automobile being serviced?
- Is the name on the contract the same as the customer's name?
- Is the service to be performed covered by the contract?
- If so, are the parts and fluids needed to service the automobile also covered by the contract?
- Is there an amount of money the customer must pay for the services performed? (This is typically referred to as a copayment rather than a deductible.)

If this information is verified to pertain to the maintenance contract, the service consultant can allow the service facility to perform the maintenance required.

When a customer's automobile has a problem and the service consultant or technician finds that the automobile may be repaired as specified by a manufacturer's TSB, the service consultant must verify the following information.

- Do the make, model, and year of the customer's automobile match the TSB specifications?
- Does the automobile's design (engine size, transmission type, brake system, and other technical information) match the TSB specification?
- Does the TSB "cure" match the customer's complaint?

If this information is confirmed to match that of the TSB, then the TSB procedures may help to repair the customer's vehicle. In some cases, a warranty may cover the repair cost associated with the TSB.

For campaigns/recalls, the service consultant must verify the following information to determine whether the automobile qualifies for a repair:

- Does the customer have a recall notice?
- If not, does the manufacturer's database indicate that the VIN is involved in the recall/campaign?
- Does the vehicle's year, make, model, and design characteristics (engine, transmission, etc.) match the campaign/recall requirements?
- Does the dealership parts department have the parts to perform the recall/campaign?

Warranty Contracts

Service consultants must disclose warranty information to customers very carefully. They must never assume a warranty contract will cover the cost of the repair. Therefore, service consultants must always prepare customers for this possibility and explain the terms of the contract to them. Service consultants must also never assume that customers have read or understood anything in the warranty contract.

For example, a contract may have a deductible that the customer must pay before the warranty company begins to pay for the repair. Some contracts may even specify a maximum amount of money or maximum labor rate that the warranty company will pay for a certain type of repair. The customer must be told about the terms of the contract and any charges the customer is responsible for, or a dispute with the service facility could occur.

To minimize the chances of a dispute between the service facility and the warranty company or the customer, or both, the service consultant must help the customer understand the warranty contract. The service consultant must also work effectively with warranty company representatives and follow its required procedures. To fully understand the warranty contract terms so each can be explained to the customer, the service consultant should start by carefully reading the entire warranty contract before the repair process starts.

After a repair has been determined to be covered, some warranty companies will outline specific procedures that the customer or service facility must follow. A service consultant's failure to follow the procedures or to have the customer agree to the repairs and the terms of the contract could mean that neither the service facility nor the customer will receive payment for the repairs.

When the exact procedure that is to be followed is not known or outlined in the contract, the warranty company must be called so the details of the transaction can be clarified. For example, a warranty company may require a warranty adjuster to visit the service facility to examine the automobile and approve the repair work before any work begins. Other warranty companies may require telephone authorization prior to a repair. When a warranty company requires authorization, a claim number must be obtained from it before work can begin. This number must be recorded on the repair order and final invoice. Again, if the service facility fails to follow proper procedures, the warranty company will not pay the claim.

Payment in Credits

In terms of payment, non-manufacturer warranty companies, which often allow any service facility to perform the repairs, typically issue a check 30 days after the repair is completed. Because of this delay and contract terms, such as deductibles and maximum payment provisions (maximum labor rate allowances for a given repair), service facilities may require customers to pay all non-manufacturer warranty contract repairs in full before releasing the vehicle. This allows the warranty company to reimburse the customer directly for the amount owed under the terms of the contract. Therefore, the service facility does not have to wait for payment and avoids not being paid in full for repairs because of deductibles or maximum payment provisions.

The payment of automobile repairs made under a manufacturer's warranty and extended warranty contracts is handled differently than a non-manufacturer warranty contract. First, the manufacturer's dealers are typically the only type of service facility permitted to perform the repairs and process the claims. This is because the manufacturer trains dealership employees how to complete the warranty claim paperwork and process it with the proper codes to identify the automobile, the repair operation, the dealer, and the technician. A service consultant who fails to follow all of a manufacturer's warranty claim procedures will have the warranty claim returned "unpaid."

Manufacturer warranty repair claims are often paid in "credits," not with cash or a check, by the manufacturer to the dealer. This is because under the franchise agreement, the dealership must buy parts from the manufacturer (except in some rare circumstances) and install them on the automobiles it services. Therefore, the dealership buys many of the parts, among other items such as tools, from the manufacturer. Then when a manufacturer owes the dealership service department for warranty services, the credits may be awarded and not money. This means the service department will not receive a check at the end of the warranty repair process, but rather manufacturer credits that can be exchanged for parts that are ultimately sold to customers.

The Lemon Law

The failure to repair a problem to the satisfaction of the customer can result in the customer filing for protection under the **lemon law**. Each state has a lemon law that requires vehicle manufacturers to "buy back" customer automobiles when they are not repaired properly within a time frame set in the state's lemon law. In most states, the law sets a maximum number of days that the automobile may be out of service (for example, 25 days), a maximum number of repair attempts (such as three) within a given time frame (typically 1 year), and a maximum number of miles (typically less than 12,000 miles) to qualify for lemon law action. Lemon laws are enforced when a customer reports the problem to the State Attorney General's Office and files the necessary paperwork to start the judicial process. This process may begin with a meeting between the customer and a manufacturer's representative in an effort to resolve the complaint. In other cases a mediator may be brought in to work with the parties, or an arbitrator may be hired to hear the facts of the case.

Because lemon law actions put the manufacturer, and in some cases the dealer, at risk, the service consultant must first ensure that the customer's complaint is accurately recorded. One manufacturer recommends that the words "customer claim" be used to begin every warranty-related repair complaint. In other words, instead of writing "The vehicle's brakes do not stop the car," the service consultant should write, "The customer claims the vehicle's brakes do not stop the car." Although the difference seems trivial, the first version implies that the service facility verified the problem and will make the repair. The second version implies that the customer has presented a possible problem that must be examined and verified by a trained technician. This is an important difference, because under most states' lemon laws a service facility is not responsible for repairing a problem until it is verified or duplicated by a technician.

In some instances, the defect cannot be found in the automobile's system and the problem cannot be replicated by the customer or technician. Also, the manufacturer's technical assistance hotline may not report any defect specific to that automobile system. When this occurs, the customer's claim cannot be considered a failure under many states' lemon laws. In such a case, service consultants must carefully and accurately document all activities according to the provisions of their state's lemon law.

Service consultants at a dealership must also carefully monitor the number of times and days that a customer's automobile has been out of service. This may easily be overlooked if the customer's automobile has been in for the same repair over an extended period of time and the customer is given another automobile to use temporarily.

For example, assume that a service consultant at a dealership takes a warranty claim on an automobile purchased by Mr. Ronald Taylor within the past 12 months and it had been driven less than 12,000 miles. Mr. Taylor's automobile had been in for repair several times over the past 7 months to the point where the number of days it has been out of service was nearing the maximum number of days permitted by the lemon law. Upon diagnosis the technician decided to replace several parts and Mr. Taylor was given another car to use temporarily. Instead of ordering the parts needed to repair the automobile overnight, the service consultant decided to include them in the regular weekly order. This caused the automobile to be out of service for the entire week, exceeding the maximum number of days under the state's lemon law statutes. The manufacturer must now "buy back" the vehicle from Mr. Taylor.

If a dealership is at fault for the "buyback," a court can legally require it to pay the manufacturer for the cost of the automobile. In Mr. Taylor's case, the dealership would have to abide by the court order since the service consultant failed to follow the manufacturer's policy to order the parts overnight to avoid lemon law problems. As a consequence, service consultants must keep their state's lemon laws in mind or a serious financial loss to their dealership could occur.

Summary

Working with warranties, maintenance contracts, TSBs, and campaigns/recalls requires service consultants to work with a number of different legal directives related to the maintenance and repair of automobiles. In addition, these agreements between automobile manufacturers and dealerships are made more complicated by state lemon laws. Thus, in addition to keeping up-to-date on the different types of maintenance and repair contracts, service consultants must know their state's lemon law.

There are some basic rules to follow when working with the different maintenance and repair agreements. Unfortunately, it is impossible to present a set of rules and procedures that will cover the different types of contracts, because there are too many variations and differences in the laws of each state. Therefore, the recommendation is that service consultants work with their facility managers and owners to integrate the procedures commonly required by maintenance and repair contracts, as well as their state regulations, into their policy manual procedures. This will enable the service consultants and other employees to comply with the provisions of the commonly used contracts and help avoid problems associated with state laws.

Review Questions

Multiple Choice

1. Which of these is NOT needed to determine applicability of a vehicle's service contract?
   A. Current mileage
   B. Vehicle identification number
   C. In-service date (the date when the vehicle's service contract began)
   D. Production date (the date when the vehicle was built)

2. A customer will receive a letter or notification from the manufacturer for which of these actions?
   A. A technical service bulletin release
   B. A vehicle campaign
   C. The end of the vehicle warranty period
   D. A vehicle recall

3. Which of the following DOES NOT describe a purpose of a technical service bulletin used by a technician?
   A. A TSB is a document mailed to customers to let them know about a problem with their vehicle.
   B. A TSB outlines how to install a redesigned version of a component.
   C. A TSB may revise a shop manual procedure or provide additional details the technician needs.
   D. A TSB may provide a repair procedure for a pattern of failures found in a vehicle or group of vehicles.

4. There are four types of automobile warranties that service facilities and their service consultants should be able to recognize and administer. Which of the following is NOT one of them?
   A. A new automobile warranty contract
   B. Bumper-to-bumper warranties
   C. Collision repair warranties
   D. Emission warranties

5. Service consultant A says a new automobile manufacturer warranty contract is almost always used by owners who buy the automobile secondhand (used). Service consultant B says the warranty coverage period pertains to a predetermined time frame and a predetermined mileage as provided in the warranty contract.
   Who is correct?
   A. A only
   B. B only
   C. Both A and B
   D. Neither A nor B

Short Answer Questions

1. What are different ways a service consultant can locate reference information about a customer's warranty, maintenance contract, and campaign or recall?
2. What is the difference between a warranty policy and a warranty procedure?
3. What are the differences among warranties, maintenance contracts, technical service bulletins, and campaigns/recalls?
4. How can a service consultant explain the differences among warranty, maintenance contract, technical service bulletin, and campaign/recall procedures to a customer?

PART I

CLINICAL PRACTICUM EXERCISE

After reading Chapters 1 through 3, visit an automotive repair business in your area known to be a leader in customer service. Once you arrive, explain that you are learning about the automotive service industry and ask if they would be willing to talk to you. Answer the following questions based either on your observations or short discussions with some of the key employees.

1. What do customers see when they first arrive, and, in your opinion, is it professional?
2. How were the employees dressed, and, in your opinion, was it professional?
3. What was the service consultant's opinion about his or her customers?
4. How did the service consultant address you when you first arrived?
5. How did the service consultant talk to customers both on the phone and in person?
6. List what makes the automotive service facility you visited a leader in customer service.

Small Group Breakout Exercises

You plan to open a service facility with your group members. Decide what type of work you plan to sell to customers and list major equipment you will need. Then choose a state in which you want to open your business, and use the Internet to determine what licenses you will need to conduct business. Also discuss what organizations you might belong to and what licenses/certifications (voluntary and mandated) you want your technicians to hold.

Explore the World Wide Web

You are the owner of a service facility. Search for automotive-related service organizations that your garage could join. Also find organizations that your technicians can join and Web sites they might use to obtain technical information. Use the Web to find out about a particular state's lemon law. Also search the Web for details about manufacturer warranties and automobile recalls/campaigns. Finally, use the Web to find warranty companies that sell warranty contracts to used car buyers and report what you find (cost, coverage, and restrictions).

PART II

COMMUNICATIONS: CUSTOMER RELATIONS

CHAPTER 6

TELEPHONE COMMUNICATIONS

─────────────

OBJECTIVES

Upon reading this chapter, you should be able to:

- *Demonstrate the fundamentals of proper telephone skills (Task A.1.1).*

- *Demonstrate how to obtain and document vehicle information and confirm its accuracy over the telephone (Task A.1.2).*

- *Demonstrate how to identify and document customer concerns and requests via a telephone conversation (Task A.1.3).*

- *Outline the steps followed in a computerized invoice system to open a repair order before the customer comes to the shop (Task A.1.5).*

Introduction

For many customers the first contact with a service facility is by phone (see Figure 6-1). When service consultants receive a customer's call, they must be aware that the reason for the call is to have his or her automobile repaired or to have maintenance work performed on it. In both cases, the customer is probably not in a pleasant mood and may be anxious, tense, and even angry. This is because for most people a trip to the service facility means they must spend money and, in their mind, will not receive anything in return such as a new coat, an entertaining experience, or a meal. As a result, the initial phone call from a customer can be the most difficult part of a service consultant's job. At the same time, it is one of the most important. How the service consultant handles this phone call is critical to the business. If the call is handled successfully, it will provide the work needed by the technicians and a sale needed for the business to be profitable.

When a customer calls, a warm welcome is difficult to convey, especially if the customer has never met the service consultant or has not been to the facility. Visual cues to show that a service consultant understands a customer's frustration and sympathizes with him or her are not possible over the phone. Of course, the service consultant must also handle other types of phone conversations effectively, such as informing customers of maintenance services, telling them the results of a diagnosis, obtaining approval for a repair order, and informing them of the charges on an invoice.

FIGURE 6-1 A service consultant on the phone with a customer.

Describing the different types of phone conversations with customers is not possible. Therefore, the purpose of this chapter is to review some basic telephone techniques. Specifically, the chapter focuses on the collection of information required to meet the customer's needs. This chapter, however, does not review the practices recommended for a telephone receptionist, although some of these would be applicable to a service consultant's job.

Basic Telephone Techniques

All service consultants should practice several basic telephone courtesies. First, the customer making the phone call should receive the service consultant's full attention. In other words, the service consultant should not attempt to do two or three other tasks while listening to a customer on the phone. Remember that with the sensitivity of the phone devices today, customers can hear background noises such as another conversation or comments by another person, the clicking of computer keys, the printing of a document, and so on. If customers have to repeat themselves because of lack of attention to their conversation, they will likely be offended.

The service consultant must attempt to put telephone customers at ease. Using a warm and cheerful voice with a professional tone is often well received by a caller. In addition, the service consultant should be upbeat with a pace that is not too slow or too fast. Remember, the reason customers typically call an automobile service facility is because they have a problem. Furthermore, they are often worried about the amount of money they must spend. In most cases, the repair of an automobile is not an option, so service consultants must be serious and professional.

Often customers call when the shop opens in the morning. Service consultants should not put a customer on hold unless they are talking to another customer. The customer who is at the facility and talking to the service consultant should be the primary focus, while the person on the phone is secondary. To learn how to put callers on hold without offending them takes practice.

Service consultants should not use technical jargon commonly used by technicians. They should use terms that a person without a technical background can understand. For example, a technician may talk about using a four-gas analyzer to measure carbon monoxide tailpipe emissions for the purpose of adjusting the carburetor. To explain this to a customer, the person could be told that the engine's fuel-to-air mixture is being adjusted with the use of exhaust measuring equipment.

When addressing customers, a service consultant should refer to them as Dr., Mr., Mrs., or Ms. When customers wish to be addressed otherwise, such as by their first name, they will request it. Of course, this would not apply to a close friend; however, when addressing a friend while other customers are present, the service consultant should be respectful.

Phone Scripts and Responses

A successful phone welcome and proper conversation require practice. New service consultants at some facilities should follow a script when answering the phone. At Renrag Auto Repair, the script and a list of responses to common questions or problems were helpful to new service consultants. The responses proved to be helpful when service consultants became flustered and were grasping for a proper reply to either a question or a problem.

An example of a script response when answering the phone is as follows:

> "Thank you for calling (name of the business)."
> "This is the service consultant (state your first and last name)."
> "How may we help you?"

First, the service consultant states the name of the facility. This tells customers that they dialed the correct phone number, assuming they are calling the facility. Second, the first and last name of the service consultant is given. When only the first name is given, the caller could become confused if another person at the facility has the same first name. Finally, the consultant says "we" instead of "I" when asking how the customer may be helped. The use of "I" could cause a caller to think that the service consultant works alone and not in a facility with more than one employee.

Some phone welcome scripts are longer and may contain a "special offer" message. For example, instead of saying, "How may I help you?" the service consultant may say, "Are you calling to schedule our oil change and free inspection special?" A common practice at some businesses is the use of a recorded message to answer the phone before a person picks up the receiver. The recorded message usually states the name of the business, any specials being offered, and then connects to the person or department being called. However, care should be taken because too much information can overwhelm or irritate a customer with a problem.

When a service facility has an advertised special on a maintenance service, such as an antifreeze flush or a tire sale, a copy of the advertisement must be near the service consultant's telephone. Important information about the sale should be highlighted, such as the date the sale ends, the cost, and any limitations or special considerations.

Obtaining Automobile Information to Start a Repair Order

Most initial calls require service consultants to make an appointment for maintenance or a diagnosis. This is when the service consultant should start to create a record of the customer in the database. As discussed in the previous chapters, the service consultant must first enter the customer's name, address, and phone number. Next, most computerized invoice systems require that the service consultant enter the automobile information before an appointment can be scheduled. The initial information required by many computerized systems includes at a minimum:

- The customer's first and last name (have the customer spell his or her name to ensure accuracy)
- The customer's address and phone number(s)
- The automobile's make, model, and year
- The services requested

Many computerized invoice systems prompt the service consultant for any additional vehicle information required, such as the VIN and mileage.

Because many customers do not have access to some of the information needed (such as the VIN), many systems allow the service consultant to bypass these entries until a later time. When a bypass is not permitted by the computer program, a series of "1's" may be entered until the information is available. In rushed or fast moving conversations, a bypass may be a good practice since it permits the service consultant to focus on the reason for the customer's call, such as an appointment for maintenance or repair of the automobile. The service consultant can obtain any remaining information after the customer arrives.

Confirmation of Customer Information

After the customer and automobile information are obtained on the telephone, the service consultant should read it back to the customer. A good practice is to again confirm the information after the customer arrives at the facility. Just as important, however, is the confirmation of the reason for the appointment.

One system we worked with requires a complete VIN and mileage before any further information can be entered. In this case, the service consultant (not the customer) must go to the automobile to write down the information. Once an accurate customer record in the database has been created, customer requests can be entered. Naturally, once the database is completed it can be quickly accessed each time the customer comes to the service facility, needs to change a scheduled appointment, or wants to schedule additional appointments (assuming the computerized system has computerized scheduling).

Phone Shoppers

In some cases, a person will call for a repair estimate, repair rates, or other price-related information. Caution, however, is required when responding to these requests. Service consultants must be aware that their competitors may be doing some "comparison shopping." The purpose of these calls is to determine what a facility charges, such as maintenance service rates, labor rates, and so on. The competitor can then set its rates and charges below those of your service facility in order to get its customers. The safest practice is to never give cost or estimate information over the phone. Only the cost of advertised and maintenance package specials should be quoted on the phone.

Other callers attempt to get a "diagnosis of their problem" over the phone, often as an attempt to fix their own automobile or to compare an estimate received from another facility. *Never, never* try to diagnose a customer's automobile over the phone or provide an estimated cost for a repair without seeing and diagnosing the automobile first. The problems that occur from such assistance, no matter how well it is intended, are too numerous to mention and also can be harmful to the reputation of the facility and the service consultant.

Identification of Customer Concerns

As mentioned previously, the service consultant must attempt to identify the service desired by customers when they call. A telephone call received for maintenance work may seem to be a rather simple request to handle; however, service consultants must not assume that all customers know all of the maintenance work that should be performed on their automobile. This is not to suggest that service consultants attempt to push customers into having work done on their automobile. Rather they should educate their customers that regular maintenance will ensure that their automobiles are in top condition, it can maximize the automobile's useful life, and it will often help the automobile's resale value.

For example, at Renrag Auto Repair, after a number of regular customers were entered into the computer database, the service consultants began to advise them of the maintenance checks and work that should be conducted on their automobiles at periodic time and mileage intervals. For instance, most automobile owners know that the oil should be changed in their automobile but forget about the need to change the antifreeze. Using the manufacturers' recommendations for the automobile was extremely helpful and kept on hand to show customers. In most cases, the customers were not aware of the manufacturers' recommendations and appreciated the information.

When customers followed the manufacturer's recommended maintenance program, the results of the checks and the work performed were entered into the computer database for the automobile. When the customer called for maintenance work, the record for the automobile was retrieved and the service consultant reviewed it with the customer and discussed any future recommended maintenance work to be performed. Naturally, the recommendations would again be reviewed after arrival.

With respect to repairs, a service consultant will receive a phone call from a person who is having a problem and needs to have his or her automobile diagnosed and repaired. This requires the service consultant to obtain information without committing the facility to any contractual promises. In this case, the service consultant must ask questions to determine the seriousness of the problem. Most importantly, the service consultant may have to cautiously advise the customer about driving the automobile to the facility. The service consultant must also be truthful about whether the automobile will be at the facility for an extended period of time and how long the customer may have to wait for the diagnosis and repair. In some cases, the repair may not take long, but obtaining the parts may take considerable time.

Obtaining detailed information about a problem over the phone from someone who is not knowledgeable about automobiles is difficult. The worst scenario in such a conversation is to permit it to progress to such a state that the customer becomes frustrated from trying to describe the symptoms. This is why some scripted responses can be helpful. For example, several scripted questions for customers with problems could be "Can you tell me when the problem occurs, such as in the morning, after the car is first started, or after it is moving?" "Can you describe the sound, such as a grinding or bumping sound?" In some cases, a service consultant can put the customer on hold to talk to a knowledgeable technician.

For example, in one case a customer was hearing a strange noise coming from the front of a particular make and model of automobile. The customer did not think it was a problem but was concerned about the noise. When a technician was asked about the noise, he explained that the automobile should be brought into the facility immediately for a diagnosis. The technician's reason was that he had recently made an expensive repair on the same make and model that was making the same noise but whose owner did not bring the automobile to the facility until it broke. The technician was correct and the customer was spared an expensive repair.

Telephone Approval to Open a Repair Order

A telephone approval for a repair is always a concern. Will the customer remember the amount of the repair? Is the person on the phone really the owner of the automobile? If a person other than the owner approved the repair, will the owner honor the approval? If not, what is the position taken by the state law about contracts?

First, all customers must be given an estimate for the cost of a repair. In some cases a diagnosis may be needed and a fee for it may be charged. The customer must also always approve the diagnosis fee either in person or over the phone. For example, if an automobile is dropped off by a customer the night before and has a problem more serious than first assumed, a diagnosis may be required.

Some diagnostic work takes considerable time and even materials to perform. In some cases, parts may need to be removed and the repair may be limited to putting on a new part. As a result, some service facilities may not charge for the diagnosis (removing the part) because it is included in the charge to make the repair. When a customer does not approve because he can put the part on himself, the person is only charged the diagnosis fee.

When a diagnosis fee or repair order cannot be signed by the customer and the approval has to be taken over the phone, the service consultant should make clear and detailed notes on the repair order as follows:

• date
• time
• name of person authorizing the repair
• work approved
• total amount of the repair the owner agreed to pay
• signature of the service consultant

When receiving approval for the work to be conducted (whether by phone or with the customer present), the service consultant should read all of the items (parts, labor, supplies, taxes, and total amount) on the repair order for which a charge is to be made. As each is read to the customer, the service consultant should make a checkmark. When the item is approved, the consultant should indicate it by placing an "OK" after the amount to be charged for the item. If an item is not approved, a line should be drawn through the item and the amount to be charged.

Summary

Working with the telephone is often an uncomfortable task for new service consultants. While experience is the only answer to becoming more skillful in the use of the phones, service consultants must be methodical. They must also be accurate when taking information on new customers and their automobiles, confirming the information taken from a customer, identifying and documenting customer requests for maintenance or repair work, scheduling appointments, and opening a repair order.

One of the most irritating phone conversations for a service consultant to experience occurs when a customer must be advised that an estimate cannot be provided without a technician's diagnosis. This sometimes occurs because of articles written by consumer advocates that automobile owners should call several automobile service facilities for an estimate of a repair and then to select the cheapest one. Because not all automobile repairs are the same even if they are for the same year and model, the best recommendation is for the service consultant to attempt to explain the reasons for a diagnosis as well as any state and federal consumer protection laws.

As discussed earlier, an estimate may be prepared after a diagnosis. An approval for a repair order must then be obtained. Although the service consultant might feel comfortable taking an approval for a repair over the telephone, the service consultant should request customers to come to the facility to discuss the repair and to sign the repair order. When a repair will cost a considerable amount of work and money (such as the replacement of an engine), the service consultant should review the work and parts needed in person and request a deposit (such as 50 to 60% of the repair charge). The service consultant might bring this up to the customer by stating that, "To get the job started, the facility requires a deposit of $_____. I will need your signed approval on the repair order along with the payment."

Finally, with the increased use of e-mail and fax, automobile service facilities should use these methods of communication to get approval for repair orders. With respect to e-mail transmissions, the computer software should permit an estimate to be attached to an e-mail message. The customer may then receive the attached estimate, review and approve the estimate or select items on the estimate, and send the estimate back as an attachment to an e-mail message stating the wishes of the owner. The e-mail would be printed and placed in the customer's file. The e-mail address would serve as the approval when used with a signature statement.

Review Questions

Multiple Choice

1. A customer enters the service area while the service consultant is on the telephone with another customer. Which of these should the service consultant do?
   A. Finish the conversation with the telephone customer first.
   B. Place the telephone customer on hold to take care of the walk-in customer.
   C. Acknowledge the walk-in customer with a wave and finish with the telephone customer.
   D. Ask the telephone customer if he or she will call back.

2. A customer recites a list of symptoms to the service consultant. What should the service consultant do next?
   A. Try to write down everything the customer says exactly the way the customer says it.
   B. Try to give an estimate of the repair cost over the phone.
   C. Try to offer suggestions about what the problem might be (over-the-phone diagnosis).
   D. Ask "open-ended" questions to try to narrow down the customer's problems and establish a priority of what he or she needs so the service facility can be of greater assistance.

3. When a service consultant is recording customer information into the database, what is one of the first things to be entered?
   A. The complete VIN
   B. The customer's name
   C. The "prime" item (reason why the customer is coming to the service facility)
   D. The vehicle's license number

4. What are some concerns about an approval for a repair taken over the telephone?
   A. Will the customer remember the amount of the repair so that when it comes time to pay he or she remembers the cost?
   B. Is the person on the phone really the owner of the automobile?
   C. If a person other than the owner approved the repair, will the owner honor the approval?
   D. There are no concerns with over-the-phone approvals.

5. If a repair order cannot be signed by the customer and the approval has to be taken over the phone, which of the following does not need to appear as detailed notes on the repair order?
   A. Date and time
   B. Name of the person authorizing the repair
   C. Work the person approved and the amount
   D. How the customer plans to pay the bill

Short Answer Questions

1. List the fundamentals of proper telephone skills.
2. How can a service consultant obtain and document vehicle information and confirm its accuracy over the telephone?
3. How can a service consultant identify and document customer concerns and requests via a telephone conversation?
4. What steps must a service consultant follow to start (and open) a computerized repair order before the customer comes to the shop?

CHAPTER 7

PERSONAL COMMUNICATIONS: FROM THE GREETING TO THE PRESENTATION OF THE INVOICE

OBJECTIVES

Upon reading this chapter, you should be able to:

- *Demonstrate appropriate greeting skills (Task A.1.6).*

- *Describe how to obtain and document customer contact information (Task A.1.4).*

- *Demonstrate how to obtain and document pertinent automobile information and confirm its accuracy (Task A.1.2).*

- *Identify and document customer concerns and requests (Task A.1.3).*
  - *Address the customer's concerns (Task C.3).*

- *Explain why repair authorization is important (Task A.1.12).*

- *Demonstrate how to open a repair order and confirm accuracy for both computerized and paper repair orders (Task A.1.5).*
  - *Use available shop management systems, computerized and manual (Task D.2).*

Introduction

One of the important qualifications for a service consultant is a genuine desire to help people. The demonstration of this desire begins with a warm welcome to the customers at a service facility. However, to be effective, the warm welcome must be appropriate.

For example, it was Jerry's first day on the job as a service consultant. Jerry was excited and wanted to show the customers his interest in helping them. Unfortunately, he was too enthusiastic when the first customer of the day came to the facility. When the customer drove up, Jerry rushed out to greet her. In his excitement, Jerry walked over to the car as the lady was getting out and in a loud voice blurted, "I am the service consultant. What is wrong with your car?" The customer, who was startled and appeared to want to climb back into her car, looked at Jerry and asked, "Where is West Third Street?" Jerry was disappointed but cheerfully directed her toward the right part of town. In this situation, the service consultant was too enthusiastic and he caused a potential customer to feel uncomfortable.

Good intentions, therefore, are not enough for a service consultant to do a good job. How customers are greeted, what is said, how it is said, and how the service consultant acts are important to setting up an effective working relationship with them.

In contrast, Tom was a service consultant and was not interested in the idea of helping people. As a result, when customers entered the service facility through the customer's entrance, they had to wait and sometimes even search for Tom. Just the sight of a customer caused Tom to get on the phone, start to file papers, or just leave the service desk to go to the bathroom or shop area. Tom was not concerned that the service facility technicians counted on him to get the work they needed to earn a living.

When customers happened to find Tom, he never greeted them or even said, "Hello." Rather, in a a gruff voice he would ask them, "Why is your car here? What's wrong with it?" The customer would describe his or her concern to Tom, who looked at the customer with a blank face. At any opportunity Tom might leave a customer in the middle of a sentence if he could find an excuse.

The descriptions of the two service consultants point out extremes in performances. Jerry was too eager and disorganized so he missed opportunities to develop an effective working relationship with the customers. Tom really did not want to greet or even help customers so he could not transition from a greeting to a working relationship with them. Jerry could be taught how to effectively greet customers and to work with them. Tom, unfortunately, does not have the important qualification for a service consultant, which is a genuine desire to help people.

The purpose of this chapter is to discuss the skill and practical knowledge needed to move from a friendly greeting of customers to establishing a professional working relationship. This positive relationship should permit service consultants to have an acceptable business conversation with customers. For example, it should set the stage for the service consultant to obtain personal customer information for the computer database, ask about the problems with their automobile, explain the need to conduct a thorough diagnosis for an accurate estimate, get an approval for a repair, collect the payment of an invoice, and have the opportunity to provide them with additional services.

Greeting New Customers

Greeting a new customer begins with a disciplined, yet warm, welcome as soon as a customer enters the service facility. Although service consultants typically greet customers, all employees should be taught how to welcome and interact with customers. The warm welcome to be used by service consultants is fairly simple but requires practice to perfect.

A warm welcome should begin with a smile and an extension of the arm to shake the customer's hand. The handshake should be a firm grip that is not too tight or too loose (see Figure 7-1). The arm should then be moved shallowly up and down two or three times before releasing the grip on the customer's hand to signify it is over.

New service consultants should practice shaking hands with another person by extending the arm and at the same time saying,

"Welcome to (the name of the automotive repair business)." "I am the service consultant and my name is _____."

FIGURE 7-1 The handshake is important to welcome a business transaction.

"Your name is _____?"

After the customer states his or her name, then the service consultant must *remember it*. Next, the service consultant should repeat the name and say:

"Mr./Mrs./Miss/Ms. (customer's name), how may we help you?"

When welcoming someone to the facility, the service consultant should maintain good (but not piercing) eye contact. The customer's response will indicate how the service consultant may be of assistance. In most cases, the customer will either request maintenance work or discuss the need for a repair to his or her automobile. This, in turn, will lead the service consultant to create a record for the customer in the computer database.

Greeting Repeat Customers

When a repeat customer enters a facility, the service consultant should offer the same warm welcome to him or her. This includes a handshake and greeting such as

"Welcome back to (name of the automobile facility)."

"I am (name) the service consultant."

When a facility has a large number of regular customers, the service consultant is not likely to remember all of their names. The greeting may then require an apology for not knowing a person's name. For example, the service consultant should say: "I am sorry but I do not remember your last name."

Documentation and Confirmation
of Customer Contact Information

As discussed in Chapter 6, when a customer is new to the facility, the service consultant must obtain his or her name, address, and phone number(s) to enter into the computer database. If the customer gave this information over the phone, the service consultant should confirm that it is accurate after the customer arrives at the facility. When a repeat customer comes to the facility, the service consultant should ensure that the information in the database has not changed since the customer's last visit.

Documentation and Confirmation
of Automobile Information

After the service consultant enters or confirms the customer contact information in the database, information on the customer's automobile is either obtained or confirmed. Most computerized management information systems require at least the make, model, year, **vehicle identification number (VIN)**, and mileage. In some cases, the computer program will permit additional information, such as the month the state inspection is due and the engine size, among other technical data.

To obtain this information for the database, the service consultant must walk to the automobile and get the information needed. For example, the VIN must come from either the VIN plate under the windshield, the tag on the driver's door, or the state's vehicle registration card. One of the most important pieces of information collected by the service consultant is the automobile's mileage for the database. This information must be accurate and must appear on the repair order as well as on the invoice so the guidelines of the state consumer protection laws, as well as warranty contract requirements, are met.

Even though the customer's automobile is entered into the database, the service consultant must review the information to ensure it is correct. For example, when a customer trades in an older automobile or purchases a new automobile, the old one must be deleted and the new one entered.

Documentation of Customer Concerns and Requests

After the database has been created and the accuracy of the information confirmed, the service consultant must focus on the reason for the customer's visit. When a visit follows a telephone conversation, the service consultant must review the information collected when the customer scheduled the appointment. When a customer comes to the facility and has not telephoned first, the service consultant must obtain and record the service request at that time.

Whether the service request is taken over the phone or in person, the first step is to classify the customer's request into one of three categories: maintenance, repair, or diagnosis. It is common for a customer to need two of these services, such as a diagnosis and repair or a repair and maintenance. The major benefit of the classification of the work to be performed is that it allows the communication between the service consultant and the customer to be more precise, and in some cases helps to schedule the work for the technicians on duty.

Maintenance Requests

Many computerized management systems permit the service facility to store different pre-priced maintenance jobs. For example, an oil change with filter may be stored in the computer for pre-priced maintenance sales at $23.95 Assume that Mr. Hudson comes into the facility and requests an oil change. The service consultant should ask if wishes to have the $23.95 special that includes the oil and filter. If Mr. Hudson agrees to purchase the special, the service consultant prints out a repair order for the pre-priced oil change. Mr. Hudson then signs the repair order to indicate his approval of the contract. After the oil and filter have been changed, Mr. Hudson pays the invoice for $23.95 plus the sales tax.

Therefore, pre-priced maintenance jobs stored in the computer save considerable time. The consultant, however, must be careful and must describe the contents of all pre-priced sales. For example, Mr. Hudson might tell the service consultant that he wishes to purchase the pre-priced special but then indicates that he prefers a specific brand of oil. Because the pre-priced special is probably based on the purchase of a large quantity of oil to reduce its cost, Mr. Hudson's request may not be possible. Mr. Hudson must be told about the brand, weight, and quality of the oil used in the special so that he understands and feels comfortable about using it in his automobile. The point is that the information on all products and their guarantees, especially those sold in pre-priced specials, must be available to the service consultant. It is also appropriate to explain to the customer that the product, in this case the oil, was purchased at a reduced price and the savings are passed on to the customers.

Repair Concerns

Some repairs, such as brake lining replacement, may not require diagnosis. They are also stored in the computer as pre-priced jobs. These repairs may not be presented as "specials," but they should be described as "package offers," meaning that the brand or quality of brakes to be installed on the car is predetermined.

As in the above maintenance special, service consultants save a lot of time entering the job into the computer because they do not have to look up the cost of the parts, calculate the markup, and then calculate the total cost for an estimate. Service consultants, however, must have the information on the parts to be sold, such as the brand and product warranties. As in the sale of the oil in a previous example, the service consultant should explain to customers that the brakes are purchased at a reduced cost by the facility so that savings can be passed on to them.

Diagnosis and Repair

Regardless of the repair to be made, even the pre-priced package repairs, a technician should examine the automobile before an estimate is prepared for the customer. The reason is the automobile may need unique parts or the problem to repair may be more complex than first thought. To assist the technician in the examination of the automobile, the service consultant must be careful to confirm the symptoms with the customer and verify that the problem is accurately stated on the repair order.

If there is a charge for the diagnosis, service consultants must disclose it to the customers so they can approve it on a repair order. As explained in Chapter 6, a diagnosis charge may be necessary because a customer may decide to have the work done elsewhere or at a later date. This charge protects the service facility from taking a loss on the time spent making the diagnosis.

In addition, diagnostic charges may be necessary because an estimate cannot be prepared without testing and/or extensively disassembling the automobile. For example, a diagnosis may include the time needed to look up the repair or diagnostic information, the test of the automobile's systems, and the need to take apart (or remove) potentially defective components to test them. As a result, some diagnostic charges may be added to the repair charges while others may be included in the repair cost. These options depend on the repair, the nature of the work, and the policy of the service facility.

When a problem must be diagnosed before a repair can be made, there are some questions that the service consultant must ask over the telephone. Some suggested questions can be found in Chapter 6. When the customer is at the facility, additional questions (presented below) should be asked. Again, these questions should be written on a piece of paper and kept at the service consultant's workstation for quick reference and to ensure that the questions are properly stated.

The more information that is obtained about a problem, the less time the technician needs to spend on a diagnosis. As a result, the service consultant must tell a customer:

> "To help our technician diagnose your problem in as short a period of time as possible and at the lowest cost to you, I would like to ask you for some additional information about your problem with the automobile."

The first question to ask is:

> "Is the automobile currently having this problem?"
>
> If the answer is *yes*, it will help the technician to know that the problem can be duplicated. The next question should be:
>
> "How long has your automobile been experiencing this problem?"
>
> Knowing whether or not the problem has just started or has been occurring for a long period of time helps the technician with several tasks, such as looking for worn or broken parts.

If the answer to the first question is *no*, then the service consultant must ask the next series of questions:

> "How long ago did this problem first occur?"
>
> "How long ago did the problem last occur?"
>
> "Does the problem occur frequently, such as every time you drive your automobile, or infrequently?"
>
> If the problem is infrequent, then the consultant should ask:
>
> "How long would you estimate is the length of time between these occurrences?"
>
> "When was the last time the problem occurred?"

Additional questions the service consultant might ask then are:

"Does the problem occur when the engine is hot or cold?"
    "What was the weather like the day of the occurrence?"
    "Does the engine turn over and not start?"
    "How much fuel was in the gas tank when the problem occurred?"
    "How do you get your vehicle (started or fixed or stopped) when the problem occurs?"

Naturally, the customer's answers to these questions dictate the next questions to ask. After the questions have been asked, the service consultant should restate the problem and summarize the customer's responses.

For example, Dr. Marks said his Aerostar stalled every morning when it was cold. After further questioning, the service consultant found that Dr. and Mrs. Marks could only keep it running if they pushed down on the accelerator pedal. The service consultant then should rephrase the information and state, "So that I have this correct, Dr. Marks, when your engine is cold, it stalls and will not continue to run unless you push down on the gas pedal? Is that correct?" After Dr. Marks agrees the statement is accurate, the service consultant should record the customer's answer into the computer.

Sometimes, after a comprehensive diagnosis, a problem may require more diagnostic time than originally agreed in the repair order. For this reason customers must be advised that the cost of a diagnosis is based on an estimated time and that this estimate is a "minimum charge." If more time is required, the customer must be called and told what the technician found, what needs to be checked further, and the possible repairs that may be required to fix the problem. The service consultant must then inform the customer that additional diagnostic time is needed and what the charge is.

Verbal Authorization

When verbal authorization for additional diagnostic time or repair is required and the authorization is given over the phone, most state laws have specific guidelines for the service consultant to follow. Because service facilities must comply with their state and local laws, they should refer to these laws and prepare a statement to be printed on their repair orders. For example, when taking a verbal authorization over the phone, many states require service consultants to write the date, time, person's name who gave the authorization, and what was said on the repair order.

Writing an Estimate

The service consultant should prepare an estimate for all maintenance, repair, and diagnostic work and review the charges with the customer before a repair order is prepared. Before the service consultant presents the customer with the estimate for a repair, a slight margin should be added to the price of parts and labor. This will ensure that costs will cover the charges and, hopefully, allow the invoice to be less than the estimate.

For example, a $150 estimate may be multiplied by 1.10 to 1.15 (such as $1.10 \times \$150.00$ and $1.15 \times \$150.00$) to get a revised estimated amount ($165.00 to $172.50) for the customer. The additional amount ($15.00 to $22.50) will hopefully cover any unexpected expense (such as an increase in the cost of a part for a particular model of automobile), miscalculations (such as a slight mistake in the labor hours for the job), and complications encountered during the repair process (such as the time to remove rusted bolts or broken parts) and even the states sales tax if applicable. Of course, if problems do not occur, the invoice would be reduced from the $165 to $172.50 estimate to $150 plus the states sales tax (if the sales tax is 6%, the total bill would be $150 \times 1.06 = \$159$). When the bill is less than expected, the customer would then be pleased!

The service consultant should remember that the information on a repair order will be read by the technician and, in some cases, a warranty auditor (if the repair is to be paid by a warranty claim). Therefore, when a customer's claim has not been confirmed by a diagnosis, vehicle manufacturers suggest that *customer claims* be written before a description of the customer's problem. For example, "The customer claims the engine will crank but won't start until the car sits overnight."

An estimate for diagnostic charges is difficult because customers want to know how much it will cost to fix, and not diagnose, a problem. In some cases, a repair labor guide may have a diagnostic charge for certain systems; for example, to check the starting system is reported in some labor guides as a one-half hour job. This means the fee charged to the customer will be equal to half of the hourly labor rate (if the hourly rate was $80 per hour, the fee to check the starting system would be $40 or half of the hourly rate). Diagnostic charges for other concerns, such as drivability problems, emission failures, or electrical problems, do not have a set fee because the verification and diagnosis of the problem can take anywhere from a few minutes to several hours. Therefore, to ensure that the technician has enough time to diagnose most concerns, a minimum fee of one and a half (1.5) hours of labor may be a standard charge. In this case the charge would be $120 ($80 per hour $\times$ 1.5 hours = $120).

Presenting the Estimate

When presenting an estimate for approval, the service consultant must disclose all costs, including the diagnostic fee, to the customer. To ensure that the customer understands the charges, a script followed by the service manager might begin as follows: "(Customer's name), in order to repair your automobile, we must begin with a diagnosis to identify the exact problem. Our minimum diagnostic fee is $_____. After the technician performs the necessary tests and analyzes the results, I will call you with the findings of our diagnosis as well as an estimate for the repair. Is that agreeable with you?" In some cases, the customer will want to know what tests will be performed. As a result, the service consultant must either know the diagnostic procedures or a technician should be present to explain them.

When a customer wants routine maintenance performed, such as an oil change, many state consumer protection laws require repair orders to present the cost of both labor and parts on the repair order for the customer's approval signature.

When the parts and labor for a repair cannot be determined, an estimate cannot be prepared nor can any information be placed on the repair order for labor/parts cost. If this is the case, service consultants should follow the directives of their state law. In most of these laws customers are required to sign a repair order waiver, meaning the customer agrees to have a repair done without an estimate of the charges. In some states, the law sets the exact wording to be used in the waiver but generally allows one or more of the following options:

- The customer can choose to allow repairs to proceed up to a certain dollar amount (the dollar amount is filled in prior to signing the waiver).
- Diagnosis and/or disassembly are to be performed for a certain dollar amount (or up to a certain dollar amount), but repairs cannot proceed until the customer is notified of the repair cost. Under some state laws, if the customer does not want the repair performed, reassembly of the automobile to the condition it was in prior to disassembly is required (in some cases without additional charge to the customer).
- The customer agrees to pay all diagnostic and repair charges without any preset amount.

Repair Authorization

Repair orders are viewed as binding contracts between a customer and a service facility (see Figure 7-2). Therefore, a customer must sign a repair order with an estimated amount for the service or a waiver if exact charges cannot be determined.

Since a repair order is a contract, it signifies that a service facility has offered to make a repair for a stated amount of money. The customer's signature signifies that the service facility has been given the legal authorization to make a repair using the stated parts and labor and that the customer will pay the amount shown on the repair order.

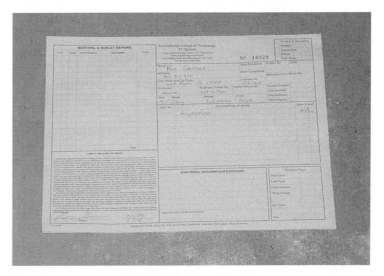

FIGURE 7-2 A repair order.

If a customer refuses or cannot pay for a repair, many states allow the service facility to hold the vehicle until payment is received. The action taken by a service facility is referred to as a **mechanic's lien**. The execution of such a lien assumes that the service facility performed the work and corrected the problem, that the repair order was signed by the customer, and that all applicable consumer protection laws were followed.

In addition, the offer and acceptance of a signed repair order means that the guidelines of the state's consumer protection laws will be honored. For example, the services performed by the service facility must meet professional standards of workmanship. If standards are not met and/or the customer is dissatisfied, then consumer protection laws permit a customer to recover the money paid for a repair.

Presenting Computerized and Manual Repair Orders

Even though most shops use a computerized management information system, some service facilities use handwritten "paper" estimates and repair orders. Paper estimates are usually written on "estimate-only" forms with the company name at the top. These forms typically have sections for the service consultant to fill in the necessary information.

When the service consultant is ready to create a repair order, a numerically sequenced form with three or four "carbon copies" is used. The repair order form often follows the same format as the one used for the estimate. When the service consultant fills in the information on the top form, it is copied to the carbon forms underneath it.

When the service consultant prepares a paper repair order, the customer's name, address, phone number, the repair request(s), parts, labor charges, and tax are written in the spaces provided. Once completed, the repair order is signed by the customer.

The bottom copy of a set of repair order forms (called the **technician's hardcopy**) is removed and given to the technician. The technician's copy is a thicker "cardboard" copy of the repair order. A time clock punches the time on the technician's form when the job is started and completed. When finished the technician should write all technical comments, such as any observations of related concerns, on the back of the hardcopy next to the time clock punches.

The technician's hardcopy is given to the service consultant, who should copy the technician's comments onto the top copy of the repair order. These comments also show on all of the carbon copies underneath the top copy. The top copy, which should show all of the charges for parts and labor, sales taxes, and the total amount owed, becomes the invoice to be paid by the customer.

When the customer pays the invoice, the service consultant or cashier writes "paid" on the top copy, date of payment, method of payment, check number if payment was made by check, and then signs his or her name to signify the transaction was completed. The top copy is then given to the customer, the second copy is used for accounting purposes, the third copy should be used to check off the parts on vendor invoices, and the technician's copy is placed in the customer's file. If a fourth copy is made, it is filed in a numerically sequenced repair order file.

The same basic process for creating estimates, repair orders, invoices, and related copies is followed in a computerized management information system. Of course, the process of copying the information from one document to the other and filing the information is far less labor intensive because the computerized database stores the information for access of different types of information in different formats.

Summary

One of the growing practices at many top service facilities is for service consultants to contact their customers after a repair to check to make sure their automobile is performing well and if the service they received was satisfactory. These follow-up calls must be made in a timely manner to maintain a positive relationship with customers.

Finally, new information technology allows service facilities numerous opportunities to save time and offer advantages to their customers, such as several pre-priced maintenance specials stored in the computer database. For example, one suggested use of new information technology is for a service facility to loan a customer a pager or obtain the customer's cell phone number while his or her car is being serviced. When the pager or cell phone rings, the customer calls the service facility as soon as possible. This enables the service consultant to talk to customers about a diagnosis, discuss a problem encountered in a repair or maintenance service, or to tell them their automobile can be picked up. This works well for service facilities near a shopping mall complex where customers can go shopping or eat at a restaurant while their automobile is being serviced.

Another practice of more progressive service facilities is to record the e-mail addresses of their customers in their database. This enables them to keep in touch via e-mail messages. Web pages also offer service facilities another excellent means to communicate with the public. In both cases, an automobile service facility can inform people about new maintenance specials, changes in technology, and manufacturer recalls, and remind them to have their automobile inspected or the antifreeze checked.

Still another use of new technology is for service facilities to use a compact disk (CD) or a video that presents illustrations of the different automobile systems and shows technicians making a repair or conducting maintenance work. These programs and videos, which are often used in vocational education programs, can inform a customer about a repair problem. This, in turn, can assist the service consultant to communicate clearly with a customer. For example, a section of a video on the brake system can be useful when explaining to a customer what a master cylinder does. Such a demonstration would assist the service consultant in meeting the expectations of ASE Task A.1.16, in which the customer is to have an understanding of the work to be performed on his or her automobile.

Review Questions

Multiple Choice

1. A customer has just given approval for repair of his or her vehicle. Service consultant A says that the technician should be provided with the approved work order. Service consultant B says documentation of the customer's approval should be on the work order. Who is right?
   A. A only
   B. B only
   C. Both A and B
   D. Neither A nor B

2. A service consultant has just completed compiling and writing up a customer's concerns. Which of these should she do next?
   A. Dispatch the work order to the technician.
   B. Arrange for a ride home for the customer.
   C. Offer an estimate for the repairs needed.
   D. Confirm the accuracy of the information with the customer.

3. Service consultant A says that when greeting customers service con-
sultants should offer their name and a handshake. Service consultant
B says that when greeting customers service consultants should make
eye contact and smile when welcoming them. Who is right?
A. A only
B. B only
C. Both A and B
D. Neither A nor B

4. A customer calls with a shopping list of problems with his vehicle.
How does the service consultant put this information in a format that
will help the technician find the customer's problem?
A. Write down everything the customer says in the order he or she
says it.
B. Ask open-ended questions regarding each item to determine the
problem.
C. Ask the customer to boil the problem down to a specific system on
the car.
D. Verify that each item on the repair order is a symptom or a maintenance request.

5. Service consultant A says that a computerized management system is
needed for the service facility to save time. Service consultant B says
the use of pre-priced maintenance jobs is needed to save time. Who is
correct?
A. A only
B. B only
C. Both A and B
D. Neither A nor B

6. A potential customer calls very concerned about an estimate received
from another shop. Which of these should the service consultant do?
A. Suggest that the other shop is probably too high and to make an
appointment.
B. Look the job up and offer an estimate.
C. Offer a discount if the customer brings the vehicle into your shop.
D. Show empathy for the customer and offer an appointment for a
second opinion.

Short Answer Questions

1. What is meant by appropriate greeting skills?
2. How should a service consultant obtain and document customer contact information?
3. How can pertinent automobile information be obtained and how can
its accuracy be confirmed?
4. How should customer concerns and requests be documented?
5. How should customer concerns be addressed?
6. Why is repair authorization important?
7. How should a repair order be opened and its accuracy confirmed?
8. What is the difference between computerized and manual shop management systems?

CHAPTER 8

WORKING OUT SERVICE DETAILS WITH CUSTOMERS

OBJECTIVES

Upon reading this chapter, you should be able to:

- *Explain why and how alternative transportation is provided (Task A.1.7).*
- *Communicate completion performances to the customer (Task A.1.11).*
  - *Identify labor operations (Task D.3).*
  - *Provide and explain estimates (Task C.1).*
- *Identify and recommend service and maintenance needs (Task A.1.10).*
  - *Describe the elements of a maintenance procedure.*
  - *Identify related maintenance procedure items.*
  - *Locate and interpret maintenance schedule information.*
  - *Communicate the value of performing related and additional services (Task C.4).*

Introduction

When working with customers, service consultants must take care of numerous details from the time of the greeting to the presentation of the invoice. They must also be ready to take care of unexpected problems and complications when automobiles are serviced. Further, service consultants must be ready to answer questions and offer advice about repairs and maintenance using a preventive maintenance schedule.

The purpose of this chapter is to discuss the details on how service consultants must be ready to take care of customers when their automobile is serviced. To explain these responsibilities, the chapter is divided into three sections. The first concerns the need to serve as a host to customers while their automobile is at the service facility. The second regards the cost and time required to make repairs and respond to customers whose estimates are not accurate. Finally, the third examines regular and preventive maintenance.

Serving as a Host to Customers

The service consultant is the "host" when customers are waiting at the service facility. Customers are inconvenienced when their automobile is being serviced. Therefore, after a service transaction, the customers' comfort and, possibly, the need for alternative transportation must be addressed.

To determine the customers' transportation needs, service consultants must ask whether the customers plan to wait at the automotive service facility, have a ride to their next destination, or prefer to have alternative transportation. Alternative transportation means that the service facility has a courtesy shuttle to take customers to their next destination, or that loaner cars or rental cars are available. Loaner cars are typically given to customers for a short period of time and are free of charge while their automobile is being serviced. Rental cars, in contrast, cost the customer money, but in some cases, a warranty company absorbs the cost.

If customers plan to stay at the service facility while service is performed, service consultants should make sure they are comfortable. Specifically, service consultants should make sure their customers have newspapers and magazines to read, know how to operate the television, and have fresh coffee to drink. In addition, service facilities should have books, toys, and activities to occupy the customers' children. Helping customers pass the time comfortably while waiting for their automobile is important but service consultants must not let this interfere with the performance of their job. For example, at a service facility in New Jersey customers can use a set of its golf clubs to play virtual golf at no charge while waiting for their automobile.

Effects on Cost and Time of Repair Completion

Customers generally have two major concerns when they enter a service facility to have their automobile repaired. The first is the cost and the second is when their automobile will be ready. Accurate estimates are important because customers do not like to be "surprised" when an invoice shows costs that are more than expected.

As noted in the last chapter, the cost estimate should be slightly higher than anticipated. The estimate, which includes labor, parts, and taxes, may be compromised because of errors in calculation, unanticipated problems when making the repair, and the cost of parts.

Service consultants must try to inform customers when their automobile will be finished. In most cases, the service consultants can estimate the amount of time a repair will take by looking it up in a labor time book (see Figure 8-1) or a computer database. To locate this information, they must be sure to have the correct year, make, and model of the automobile as well as other characteristics, such as engine size and transmission type. Unfortunately, the availability of parts and repair complications (a part is rusted and cannot be removed quickly) can extend the time it takes to make a repair. When this occurs, the customers must be notified immediately because they do not like to be "surprised" when told that their automobile is not ready when promised.

FIGURE 8-1 A labor time book.

Influence of Parts Orders on Cost

An estimate presents the prices for parts and labor to a customer. The estimate assumes that the information received on the cost of the parts to a facility is accurate.

To arrive at a price to charge a customer, the service consultant begins with the cost the service facility must pay for the part. The part is then resold to the customer at a price that is **marked up** by a prespecified amount or percentage over the cost. **Markups** are needed to cover the expenses that a facility incurs to obtain or pick up parts. This also may include the need to prepay for parts and store them in the inventory, the time the parts specialist takes to look up the parts and order them from a supplier, and the time managers or owners spend to make arrangements with parts suppliers. In addition, the markups must cover the work associated with core returns or the proper disposal of old parts in accordance with local, state, and federal laws.

The part number and price to be charged to the customer are shown on the estimate as well as on the repair order once the customer authorizes the work (see Figure 8-2). Therefore, the amount of the markup must be carefully calculated. To relieve the service consultants from calculating parts markups, many computer programs calculate them when a part or the cost of a part is entered. The price charged to the customer is then added to the estimate.

Information about the amount a service facility pays for a part is obtained from the parts supplier, the computer database for items in inventory, a parts price book, or a computer link when the service facility is linked to a parts supplier. Of course, the critical task is for service consultants to be sure that the information needed to order parts is correct. This

FIGURE 8-2 A parts manager checking parts against a repair order.

may seem to be an easy task but a part for the same year and make of automobile may be different for the various models, engine size, and so on. If an incorrect part is identified, a difference in the cost of the part to the service facility is very likely.

Influence of the Availability of Parts

The next piece of information that is critical to the service consultant is whether the part is "in stock" (called parts availability) or if it must be ordered. Customers must be told when the part is not immediately available because they may need to return the automobile at a later date for the repair.

Therefore, the availability of the parts must be confirmed before customers are told when they can pick up their automobile. The availability of a part is also needed to ensure it arrives prior to the start of the repair. For example, if a technician has removed an old part from an automobile and does not have the new part, a service bay is tied up and time is lost on the repair. This costs the service facility and even the technician money, and the customer will not have the automobile when expected. This means the owner, manager, technician, and customer will be unhappy!!

Keeping the Customer Informed

After a comprehensive estimate of the cost and the time it will take to complete the repair have been calculated, the service consultant must inform the customer. The service consultant should begin by explaining the charges to the customer by stating, for example, "According to our database, this repair will take ___ hours and the parts, labor, and tax will be less than $_____. Assuming the parts are available and the work in our shop stays on schedule, your automobile should be available by ___."

Again, the service consultant must remember that the estimate and repair order charges should be slightly inflated so the final invoice will be less than expected. Also service consultants must not "over promise" the time when the automobile will be available. The automobile repair should be completed and ready to be driven off the service facility's parking lot when promised. Therefore, a customer should never have to wait until it is taken out of the technician's bay, test driven, and cleaned up.

As noted previously, and restated here as a review, some service facilities have a policy that service consultants must add the parts and labor together and then multiply the total by 1.10 to 1.15 to obtain the total cost including sales tax. The amount to be charged to the customer should then be less than the estimate. If any minor problems or mistakes occur, such as data entry errors, incorrect parts prices, or unexpected problems in the repair process, the cost cushion should cover them. If the cushion is not enough, the customer must be called immediately to explain the problem.

When the Estimated Cushions Fail

Unfortunately, despite the service consultant's best efforts, a final invoice may exceed the estimate or the automobile will not be ready when promised. The reasons for this are often due to various factors, such as the use of ever-more-complex automobile systems. The key to handling these unexpected problems and complications is to keep the customer informed.

For example, a single defective part may have several different replacement choices, each with a different price and availability. In addition, a complex system may cause technicians to overlook related components that are defective or are on the brink of being defective when the diagnosis is conducted. These situations result in errors, which, in spite of the best efforts of the service consultant and technicians, will exceed the estimate, and the time needed to make a repair will be longer than expected.

How these differences are handled depends on service facility policy. When a technician runs into a problem, the service consultant must be made aware of the situation immediately. Then various options should be considered to rectify the problem. For example, the technician and the service consultant need to work together and redo the estimate. Hopefully, they will be able to come up with options that can be presented to the customer. Then the customer must be called to explain the situation and the possible options available.

Automobile Maintenance

The AAA estimates that the majority of automobile owners ignore the maintenance recommendations made by manufacturers. The United States federal government knows this all too well, so it empowered the EPA under the Clean Air Act to oversee the implementation of emission inspections and maintenance (I/M) programs. These programs require the emission inspection of automobiles in highly and densely populated areas and require repair if the automobiles do not pass inspection. The inspection regulations vary from state to state and even from area to area within some states.

Owners neglect the maintenance of their automobile for different reasons. For example, they may not want to spend the money, or they may be unaware of the maintenance needed to keep their automobile in sound operating condition, or they may not believe maintenance is really necessary. Service consultants must advise their customers of the maintenance services recommended for their automobiles and explain how each benefits them.

When advising customers, service consultants must explain that the maintenance and repairs needed for older and higher mileage automobiles are different from newer automobiles. The sale of maintenance packages at certain mileage intervals, for example, 15,000 miles, works best for relatively new automobiles such as those that are less than 4 years old and under 70,000 miles. These packages are often based on the manufacturer's recommended maintenance schedule, which is designed to keep the customer's automobile in top mechanical condition. Naturally the sale of a maintenance package assumes the customer with a newer vehicle wants to keep it in top condition and wants the manufacturer maintenance services done at the suggested mileage intervals.

To obtain the maintenance schedule for the customer's automobile, the service consultant can look in the owner's manual, an aftermarket labor time publication, or a computer database that lists the manufacturer's recommended maintenances by mileage. Once the labor time is known for each of the maintenance items, the service consultant can calculate the cost of the maintenances. The service consultant should not forget to add the cost of parts, fluids, and supplies.

To speed up the process, many service facilities use **pre-priced maintenance menus**. As shown in the example in Figure 8-3, such menus may list the different maintenances found in the manufacturer's manual down on one side with the mileage intervals across the top. The center of the chart shows the total price for each service with the total cost of the combined service for every mileage interval. Figure 8-3 is a miniature example of a service menu. When a customer decides to order one of the services, the service consultant will identify the service and enter it on the estimate.

To suggest these services to the customer, a service consultant may consider the following approach after addressing the customer's automobile repair needs:

"Now that we have addressed your repair needs, your automobile has ____ miles. (Note: the current mileage was already verified and entered into the database at this point.) The manufacturer of your automobile

SERVICE	MILEAGE			
	3,500	7,000	10,500	15,000
Oil Change	$15	$15	$15	$15
Lubrication		$ 4		$ 4
Tire Rotation		$ 9		$ 9
Alignment				$30
	$15	$28	$15	$58

FIGURE 8-3 A maintenance schedule.

recommends the following services at ____ miles." Next the service consultant should show the customer the maintenance menu, the computer database printout of the services, or the page from the manufacturer's/owner's manual. Then, the service consultant might offer the following to the customer, "If our technician could have the recommended services completed before you pick up your automobile, would you like for us to complete them for you?" If the customer agrees to have the maintenance done, it should be entered on the repair order.

The method described above works well for newer automobiles whose owners want to ensure that every maintenance procedure is done at the interval recommended by the manufacturer. However, service consultants may find that customers with older or high mileage automobiles are not interested in the same menu. This is mainly because these customers are more concerned with "keeping their automobile going" with the repair and maintenance of systems that prohibit function of what they consider "critical components." Therefore, these customers may be more receptive to maintenance specials packages.

Maintenance Specials

Typically, maintenance specials are for the more common maintenance services, such as an oil change. In states that require an annual state safety inspection, the inspection service can be a time when customers also receive seasonal services. A seasonal maintenance service includes such services as an antifreeze flush in fall or spring. When a seasonal maintenance is combined with other services, such as an inspection, a discount is often offered and is referred to as a maintenance special. This is often popular and has a broad customer appeal.

To suggest maintenance specials to customers, service consultants should remember that part of their job is to provide advice. The suggestion about maintenance is intended to help customers keep their automobile in top condition. One approach to make a customer aware of maintenance specials is to have the service consultant state: "Now that we have addressed your concerns, we have a special that gives you. . . (provide an appropriate service recommendation such as). . . an oil change and antifreeze flush for $_____. If our technician could have this service finished before you are ready to pick up your automobile, would you like to have that done?" If the customer agrees to have the service done record the sale on the repair order.

Service consultants should not be discouraged when customers refuse to have a recommended maintenance performed. Some customers may want to think about it until the diagnosis or inspection is completed and they know more about the cost of repairs. This is not a cause to worry because there will be a second opportunity to promote the suggested maintenance again when the service consultant calls the customer with the results of the technician's findings.

Finally, a method used to promote maintenance specials is to keep regular customers informed about the services their automobile needs when they schedule an appointment or arrive at the service facility. The information needed to provide this type of customized assistance comes from the computer database. Most computerized databases contain a wealth of information, such as past estimates given after a previous repair was completed, past maintenance services performed, and, naturally, the vehicle's age as well as past mileages. This information can be used to focus attention on critical repairs or maintenances that the customers' automobile needs. For example, if the database shows that there has been 10,000 miles driven since the customer's last tire rotation, the service consultant should suggest a tire inspection and rotation while the automobile is in the shop.

Preventive Maintenance

Preventive maintenance usually involves the inspection and replacement of parts before they break. This is an important service not often considered by automobile owners. One of the reasons these services are not considered is that many owners do not know what needs to be changed on their automobile.

For example, assume an automobile engine has 75,000 miles and the automobile manufacturer recommends a timing belt replacement at 60,000 miles. The service consultant should examine the customer's file in the database, and if the belt has not been replaced in the past it should be brought to the customer's attention. The service consultant should advise the customer that the belt should be changed and explain why.

Other items to check in a customer's file in the database include the most recent brake pad thickness readings at the last service relative to the current mileage of the automobile. If the lining was thin and the automobile was driven a lot of miles since the last check, it should be inspected again. In addition, the facility should keep track of when each customer's state safety and emission inspection is due as well as when the engine had its last oil change, the spark plugs replaced, the radiator hoses and fan belt changed, and the fuel filter replaced. Because computer programs can identify customer automobiles based on past estimates and services, some facilities send out reminder notices. These notices should remind the customer of any maintenance services needed and the specials that are being offered.

Related Services

When customers request repairs, service consultants should always consider any maintenance services that should be performed at the same time. For example, when a water pump that is driven by the timing belt is replaced, it may make sense to replace the timing belt. This will help to avoid a breakdown after the old belt is re-tensioned to the new water pump. The opposite may also hold true because old water pumps with loose bearings may seize soon after a new timing belt is installed and re-tensioned. Needless to say, old water pump bearings that seize will ruin a new belt. Therefore, the service consultant must help the customer understand the relationship between the system parts and should ask the technician to pay special attention to the condition of related parts.

Likewise, when a repair is performed, such as the replacement of a head gasket, it makes sense to change the oil, drain and replace all of the antifreeze, and maybe even replace the spark plugs at the same time. These services will reduce the likelihood of a costly comeback from oil contaminated with small particles of dirt and small amounts of antifreeze that have drained into the oil. These services also benefit the customer because he or she saves money if the work is done when the head gasket is replaced. This is because the antifreeze is drained from the engine and the spark plugs are readily accessible when the head gasket is removed. Therefore, the labor cost is minimal because most of the charge is for replacement parts (spark plugs and oil filter) and fluids (antifreeze and oil).

Summary

The cost and time needed to make a repair are difficult for service consultants to estimate. Nevertheless, customers expect accurate estimates for costs and time before they approve the service. To ensure that the work remains on target, service consultants must constantly monitor the status of the maintenance or repair of the automobiles. This requires them to be aware of the availability of parts and their delivery time, the time being spent on a repair, the workload of the technicians, and the problems that technicians encounter. Service consultants must be even more aware of the progress of a repair or maintenance for older automobiles. Because of age and rusty parts, the services for these automobiles have the potential of taking longer than normally expected.

Estimates of cost and how long an automobile will be at the service facility are as much an art as they are a science. Labor times, parts availability, the workload of the technicians, the age of the vehicle, the complexity of the system being serviced, and the experience of the technician with both the service procedure and the automobile's design are all factors to be considered. While a cushion for additional time and cost can help, there is no exact formula to use to know how much to add to an estimate for many automobiles. Rather, a service consultant's experience, luck, and ability to comprehend all of these factors all have to go into the estimate.

Review Questions

Multiple Choice

1. Service consultant A says that telling the customer when the vehicle will be ready at the time he or she drops it off creates expectations. Service consultant B says that accurate completion times can only be determined after vehicle inspection/diagnosis. Who is correct?
   A. A only
   B. B only
   C. Both A and B
   D. Neither A nor B

2. When a vehicle is found to need maintenance work that was not requested by the customer, a service consultant should recommend it because:
   A. it provides additional income for the service facility.
   B. it is the responsibility of the service facility to advise the customer of his or her vehicle needs.
   C. the customer must have this work done to maintain the vehicle's warranty.
   D. it keeps the vehicle in good working order.

3. A service consultant needs to find alternative transportation for a customer who is under 21 years old. Policy dictates that rental/loaner cars cannot be given to anyone under age 21. What do you do?
   A. Offer a reduced rate rental car.
   B. Lend the boss's vehicle.
   C. Offer directions to the nearest bus stop.
   D. Offer to provide a ride.

4. Service consultant A says that providing an estimate is required by law in most states. Service consultant B says that explaining the details of an estimate helps to add value to the services the customer is buying from the shop. Who is correct?
   A. A only
   B. B only
   C. Both A and B
   D. Neither A nor B

5. Service consultant A suggests that offering a customer a ride home or to work represents alternative transportation. Service consultant B says that driving the customer to the bus stop is providing alternative transportation. Who is correct?
   A. A only
   B. B only
   C. Both A and B
   D. Neither A nor B

Short Answer Questions

1. Explain why and how alternative transportation is provided.
2. How are completion performances communicated to the customer?
3. How are labor operations identified?
4. How should estimates be presented and explained?
5. How can a service consultant identify recommended service and maintenance needs?
6. What are the elements of a maintenance procedure?
7. How can you identify related maintenance procedure items?
8. How can maintenance schedule information be located?
9. How is the value of performing related and additional services communicated to the customer?

CHAPTER 9

CLOSING A SALE

OBJECTIVES

Upon reading this chapter, you should be able to:

- *Explain how to promote the procedures, benefits, and capabilities of the service facility (Task A.1.8).*

  - *Close the sale (Task C.7).*

  - *Explain product/service features and benefits (Task C.5).*

  - *Overcome objections (Task C.6).*

- *Identify and prioritize vehicle needs (Task C.2).*

- *Explain how to present customers with the work to be performed and related charges, and review the methods of payment (Task A.1.16).*

Introduction

One of the responsibilities of service consultants is to promote service facility sales. To accomplish this, service consultants must prescribe appropriate services that help customers solve their problems. More commonly, service consultants suggest services that help keep customers' automobiles in top condition, such as the maintenance specials discussed in Chapter 8.

When promoting sales, however, service consultants must avoid **overselling** the customers. In other words, they should not try to sell customers any services their automobile does not need. When suspected of doing this, they can lose the confidence of their customers and may even break the law.

To avoid the temptation to oversell, service consultants must put their customers first and be committed to offering recommendations that will enable their automobiles to run longer and better. To assist them, service consultants should refer to the recommendations made by the engineers who manufactured the automobile, the expert advice of the service facility technicians, and management's seasonal specials aimed at assisting customers at discounted prices.

The purpose of this chapter is to describe some of the procedures and techniques service consultants should follow when selling services to customers. First, to make a sale, service consultants must understand the products and services sold by their facilities and how they benefit customers. Next, service consultants must be able to explain clearly the benefits of the service to customers in relation to the needs of their automobile and how the work will be performed. Finally, in order for a facility to be a "full-service facility," service consultants must be able to offer and explain the different methods customers can use to pay their bill.

Closing a Sale

The previous chapters discuss the activities of the service consultant, from the greeting of customers to the presentation of the invoice plus the reasons why complications occur during this process. At any time during this process, however, customers may walk away and service consultants lose the sale. The service consultants must be mindful that one of their functions is to get customers to accept their recommendations and approve an estimate for maintenance or repair. This is called **closing a sale**.

To get the signature that closes a sale, service consultants must be genuine in their desire to help their customers. If not, they may be seen as insincere "snake oil" salesmen, meaning people who sell something that is worthless. One way to communicate sincerely is to show a willingness to assume the customers' problems and to help them make their automobiles run longer and better.

Therefore, service consultants must think of customers as clients or friends and recognize that their automobiles are important personal possessions, often viewed by them with pride. For example, for one reason or another, such as body style or gasoline efficiency, customers typically hand pick their automobiles when they purchase them. Service consultants must take care to never put down their selection of an automobile because to do so implies the customers made a wrong decision.

In addition, when treating customers as clients or friends, service consultants must explain the work to be performed on their automobile in a patient, calm, and logically prioritized manner. Then they must explain how the facility is equipped to perform the work and why the technicians are capable of making the repairs and maintenance needed on their automobile. This, of course, is easier said than done when a service facility is busy, but the techniques must still be practiced.

Telephone Customers

To get to the point of being able to close a sale with customers who call for information, the caller must be enticed to come in for service. For example, as mentioned previously, many customers seek assistance over the phone before coming to the service facility; they also try to shop for the best prices for a repair by phone. In these cases, service consultants must try to convince them to bring their automobile into the service facility for a proper review and possible diagnosis. This is often difficult because the difference between obtaining the customer's business and losing it depends on whether the service consultant is able to project an image of caring about the customer's problem. Thus, this step is important and difficult.

To convince customers who call on the phone that the service facility can and wants to help them solve their problems and relieve them of their worries about their automobile, service consultants must carefully focus on their choice of words and tone of voice. For instance, service consultants should take a positive position when letting customers know that the service facility can definitely solve their problem (assuming that is true) if they bring their automobile into the facility. With this statement and a proper tone of voice, service consultants should convey their desire and willingness to help customers and instill confidence that what they say is honest and true. For example, statements that imply doubt, such as "I think we can take care of your problem," or any negative thoughts, such as "Well, if we cannot fix your car, you can trade it in for another one," should never be used. The main idea is to tell customers they are important, their problems are important, and the service facility wants to help.

When customers who call a facility agree to bring their automobile in, service consultants may either arrange for a tow truck to pick up the automobile or provide directions to the facility. As explained in Chapter 8, when customers arrive at the facility, the service consultant must greet them properly, prepare the estimate, and then close the sale.

Selling Repairs: Prioritizing Automobile Problems

One recommended approach to working with customers to sell the repairs needed on their automobile is to use a method that prioritizes problems from the most dangerous to the least problematic. This method explains the work to be done, and the service consultant can promote the ability of the service facility to do the work. In addition, this method presents the opportunity for the service consultant to overcome any customer objections before they are stated.

One way to prioritize repairs is for a technician to examine the automobile (see Figure 9-1) and then provide a list of repairs for the service consultant to review. The technician should prioritize the list of repairs either by placing the most important first or by assigning each one a number, with the lower numbers representing a higher priority (see Figure 9-2). For example, a safety problem, such as an axle that is ready to fall off, will have a higher priority number (say a number 1) than a piece of trim that is loose (say a priority number of 10). In between these numbers,

FIGURE 9-1 Technicians making a repair.

FIGURE 9-2 Levels of repairs.

other items of varying importance would be shown, such as tire wear at the edges (say a number 4) and small oil leaks at the valve cover (say a number 8).

Another approach to this method is for the service facility to use **repair categories**. A copy of these categories should be shown to customers and posted on a bulletin board at the workstation of the service consultant. The categories are as follows:

1. *Imminent Danger*—Safety repairs that will cause harm if not taken care of immediately. An example would be steering components that are very loose and ready to break.
2. *Hazardous Danger*—Safety repairs that could cause harm if not repaired soon. An example would be a tire with cords showing.
3. *Imminent Malfunction*—Non-safety concerns that will soon cause a breakdown if not repaired. An example would be a starter motor that has excessive draw and cranks the engine slowly when cold.
4. *Potential Malfunction*—Non-safety concerns that may lead to a breakdown if not repaired. An example would be a starter motor that has higher than normal current draw.
5. *Nonessential malfunction*—Non-safety concerns that will cause a system to not operate but will not lead to a breakdown. An example would be an air-conditioning system that has a leak.

6. *Nonessential Concern*—Non-safety concerns that may cause a system not to operate as intended but will probably not lead to a malfunction. An example would be a slight oil leak at an engine valve cover.
7. *Noteworthy items*—Non-safety concerns that a technician notices but really are not important to the function of any system. An example would be a cracked trim piece that covers a component for decorative purposes only.

After an automobile is examined, the technician will determine the priority level the repair represents. Then the service consultant must report the repair and its level to the customer. This procedure ensures that the service consultant does not exaggerate or understate the seriousness of a problem, and it ensures that the customer understands the relative importance of the repair. When done properly, value is added to the customer's service visit because the examination of the automobile covered seven levels of possible problems and they were reported to the customer.

Of course, some technicians do not have either the ability or the willingness to examine a vehicle for potential problems. For example, a technician may not think the problem is important to point out because it is not something that he or she would fix on his or her own car. Another common reason may be that the technician does not want to do the repair because it is not the type of work he or she likes to perform. When a technician has this attitude or lacks the ability to identify potential problems, he or she should be given work that does not require the inspection of automobiles. Rather, after a competent and conscientious technician checks an automobile, the work may be passed on to a less capable technician.

Maintenance Sales: The Feature-Benefit Method

Convincing customers to purchase maintenance services is often difficult. As a result, the sale of maintenance services depends on the service consultant's ability to convince customers that these services will help them. One technique is for the service consultant to focus on what the recommended maintenance will provide to the customer (feature) and how the maintenance will be an advantage to the operation of the customer's automobile (benefit). This technique is referred to as **feature-benefit selling**.

The feature-benefit selling technique relies on the service consultant's knowledge of the maintenance schedules recommended by automobile manufacturers and the seasonal maintenance specials offered by the facility. Then when discussing any services needed by a customer, the service consultant attempts to sell these features, meaning what the manufacturer recommends, the seasonal specials offered by the service facility, plus any recommendations made by the technicians conducting a diagnosis or inspection. To accomplish this, the service consultant must carefully describe to the customer how a maintenance service will benefit him or her. Common benefits may be extending the life of the automobile, safety, improved operation of the vehicle, and reduced operating costs.

An example of feature-benefit selling is a recommendation for a battery service because of corrosion. The feature, of course, is that the battery terminals and cables will be cleaned. To make the service relevant to the customer, the service consultant also explains the benefits of the recommended service. In this case, the customer should be told how the battery service helps extend the life of the battery and reduces the possibility that the battery may fail to start the engine. As a result, the battery service (feature) can prevent an inconvenient and possibly costly breakdown (benefit).

Feature-benefit selling is a common technique because it does not enter into the technical details about how an automobile works but states simply and briefly the work to be done (feature). For instance, too often, a service consultant tries to impress a customer about how technically difficult or complicated a maintenance or repair job is and the customer loses interest. The service consultant then may lose the sale because the customer becomes overwhelmed by the details. Therefore, this selling approach helps simplify what the customer wants to know; specifically, how the service can help the customer and why it should be purchased (benefit).

If the benefit is understood and appreciated, customers will most likely purchase the recommended maintenance. These sales are the backbone of a service facility because they help to maintain stable sales income. In other words, a service facility can hire the best and fastest technicians, but the service consultant must make maintenance sales to keep the technicians working.

Of course, in order for service consultants to use the feature-benefit selling technique effectively, they must have something to sell. This requires a list of manufacturer recommendations for automobiles, seasonal special offers by the owners/managers, and recommendations from the technician about the automobile's condition. Therefore, when examining automobiles for repairs, technicians must also check them for possible repairs and maintenance work. This does not mean that technicians should remove any parts to identify needed services but should simply examine each customer's automobile and point out items that need repair. The inspection forms used by Renrag Auto and shown in Chapter 4 can be used for these reports. To help customers understand the inspection process it sometimes helps for the service facility to have a waiting room that over-

FIGURE 9-3 A waiting area overlooking service bays.

looks the service bays. This allows customers to watch how technicians are going about their examination (see Figure 9-3).

Four Opportunities to Sell

Closing a sale often takes persistence. However, knowing when there is an opportunity to make a sale comes before persistence. There are four opportunities for a service consultant to make a sale to a customer.

The first opportunity to sell occurs when a person calls the service facility and talks to the service consultant. The customer's name should be entered into the database and an appointment should be made. During the conversation, the service consultant can suggest maintenance specials, but the important objective is to get the person and the automobile into the facility.

The second opportunity to sell is during the initial service write-up. This opportunity allows service consultants to make customers aware of specials and to suggest maintenance services specific to their automobile. A benefit presented to the customers should include the opportunity to "keep your vehicle in top mechanical condition."

Another service to be offered during the initial write-up is to inspect the various vehicle systems. The benefit is to "provide you with information about your automobile so you can include any problems we might find into your financial decision." For example, assume Mr. Hoffman is having his vehicle diagnosed for a serious engine problem. He may want to have information about other vehicle systems, such as the brakes or air conditioning. At the very least, this information can help him make a financial decision about the overall condition of his automobile to determine if the engine is worth repairing. In addition, if he decides to have the engine repaired and the technician finds that the air conditioning does not work, Mr. Hoffman may want it fixed at the same time that the engine is being repaired.

The third opportunity to sell is when service consultants call customers for authorization to perform a repair or to inform them of when a repair or service will be completed. At this time, service consultants may ask the customer to reconsider previously suggested services or repairs that were declined during the first and second sales opportunities. This also gives the service consultants the opportunity to point out any other items the technician may have found in need of repair or replacement while working on the automobile, such as replacing a windshield wiper, replacing a bad tire, and so on.

The fourth opportunity to sell is when customers pick up their automobiles. Service consultants should present the keys to the automobile to the customers and then walk them to it, preferably near the service entrance. This is called an **active delivery**. During active delivery, service consultants have a final opportunity to "connect" with the customer and discuss any issues of importance.

In the first part of active delivery, service consultants should convey to the customers positive points about their automobile. This is suggested because during the sales process, customers have been told of all the things that are wrong with their automobile. This can have an unpleasant effect on customers about the condition of their automobile and the wisdom of the payment they have just made for repairs. In some cases, they may even feel like they are "driving junk." No one who has just paid a service facility for repairs should think he or she is leaving with junk. So service consultants should mention some of the positive features of the customers' automobile. For example, the service consultant may point out the automobile's interior, gas economy, paint job, stereo system, the attractiveness of the model of automobile, and so on. In addition, the service consultant should note how much better the automobile will perform now that the repair has been made and how the customer will not have to worry about another repair for quite a while.

In the second part of active delivery, service consultants should try to sell to the customer, one last time, any needed services that were previously declined. This can be accomplished by asking customers if they would like to schedule a future appointment to perform a previously suggested repair or to schedule the next regular maintenance, such as an oil change. In some cases, showing customers any worn or bad parts that need to be replaced in the future can make a future sale that was declined. To make this sales opportunity work, customers must be shown (not told) about their vehicle's problem(s). Even if customers again decline the repair after seeing the part, they should be impressed (again) that the service consultant is interested in helping them with their automobile problems.

While many service consultants are not opposed to active delivery, they often feel they barely have enough time to take care of their other customers who need help. Therefore, at some service facilities customers will see a cashier to pick up their automobiles. If this is the policy of the automotive service facility, then service consultants should at least call their customers after the automobile is finished. Then they should engage customers in a discussion described in the active delivery conversation, such as pointing out the positive points of their automobile, and then try to schedule additional repairs or maintenance. This is not to say that time is still not at a premium for service consultants, but it can cost the service facility sales and reduced customer satisfaction if the active delivery conversation is not attempted.

Impact of the Environment on Sales

Service facilities operate in **open business environments**. This means there are forces outside of their control that affect their sales potential. Service consultants, managers, and owners should always attempt to take advantage of the positive environmental features and counteract their negative influences. Two of the business environments important to sales are the tactical and the operational environments.

The tactical environment refers to the support arrangements needed to conduct business. For example, an automotive service facility needs to have auto parts to conduct business. The availability of one or more auto parts vendors, or suppliers, can influence the service facility's ability to make a sale. A sale is more likely if parts are immediately available than if they will not be delivered for a week. A negative influence from the tactical environment must be counteracted by internal strategies, for example, how the facility can get the part sooner.

At the same time, a facility with a supplier next door can use that to its advantage by telling customers how soon it can get the parts needed to make repairs. Also a service consultant at a facility with several parts suppliers available to it can tell customers how he or she will "shop around" three parts suppliers to get the best price.

The operational environment concerns factors related to conducting day-to-day business activities. These factors may be the local demographics, economics, and sociocultural features of a community. Obviously, a business must appeal to the people in its community and recognize their interests, preferences, economic status, and so on.

For example, people who live in a community with a very high average income may own or lease new automobiles that have comprehensive warranty contracts. Sales for repairs may be limited in such a community. On the other hand, used cars that are not under a new car warranty may be found in a town or city with residents with lower incomes. In this case, repair work may be in greater demand. Likewise, if a town factory closes and a large number of people lose their jobs, the economic environment will change and sales problems must be expected.

Service facilities should always consider ways to take advantage of or have an influence on their operating environment. One example is to open a new service facility in or near a populated area with easy access to and from a heavily traveled road. Another is to ensure that the facility's building is large enough to handle the number of automobiles needed to generate the sales needed for a profit. This would include room enough for vehicles in the process of having long-term work done on them as well as those needing short-term maintenance work. At the same time, the property (not just the building) must be clean, attractive, and landscaped (see Figure 9-4) and the interior should be clean and appealing to

FIGURE 9-4 An attractive service building.

FIGURE 9-5 A service facility in an undesirable location.

customers (not to the owner or workers). Old parts, parts that are being returned to a vendor, and parts that are delivered to a facility should never be lying around where customers can see them. Customers must always feel comfortable and never uneasy when they are at a service facility.

Of course, some features, such as location (see Figure 9-5), may not be able to be changed but some, such as cleanliness, can easily be improved. Another option that influences sales is the facility's operating image. This could be improved, for instance, by increasing advertising, offering shuttle services, promoting maintenance specials and discounts, or adding new services (such as a car wash or quick lube station).

Selling and Methods of Payment

Closing a sale often depends on the method of payment that a customer can use. Service facilities must provide customers with as many alternative methods as possible to pay their bills. Then service consultants must understand the financial options that customers can use to pay for their service. When customers need some other means to pay for a repair that will cost a lot of money, a sale may be closed because the service consultant can make the arrangements for them to finance the charges.

Therefore, when service consultants are trying to close a sale, they should be able to offer several methods of payment that a customer can

FIGURE 9-6 A service consultant explains payment methods to a customer.

use. These methods may range from short-term credit provided through the facility (such as credit provided to fleet vehicles) to long-term loan arrangements through a bank. At Renrag Auto Repair, many of the larger repair jobs, such as an engine replacement or overhaul, were sold because customers were able to use one of these methods to finance the repair (see Figure 9-6).

One of the most popular methods used to pay for automobile services is by credit card. Although this option costs the service facility money when it processes a charge and pays for an extra phone line, it is necessary for a business to be successful. Service consultants must be able to use the credit card machines plus close out at the end of the day.

For customers who do not use credit cards, service facilities must be able to accept personal checks. However, because a customer's check can "bounce," meaning the person does not have enough money in the checking account to cover the payment of the check, a facility may use a check cashing service. For a fee, this service guarantees the facility will receive payment when a check bounces. Service consultants must be sure that all checks are properly written for endorsement and deposit.

In addition, service facilities should have financial arrangements available for customers making a large purchase (such as a set of tires) or needing an expensive repair. In these cases, through arrangements with a finance institution, the service consultant could offer customers a 30-day same-as-cash option or low monthly payments. These approvals typically take one half hour or less to approve so that customers can have the work started on their automobiles right away.

Large financial loans, for example, for the replacement of an engine and transmission costing more than $3,000, can be offered by a service facility through a bank. The loan application is usually made at the service facility and then faxed or taken to the bank for approval. The bank checks the customer's credit and, in most cases, the approval is obtained before the end of the working day. When the loan is approved, the bank will send a check to the service facility to be endorsed by the customer and the service facility manager or owner. The technicians can then begin the long-term repair and parts can be ordered. To make this type of financial arrangement appeal to customers, it must be convenient and must have reasonable finance charges. Needless to say, the service consultant must carefully explain the process, loan agreement, and benefits to the customer.

Summary

While the service consultant's actions and attitudes are important to closing the sale, the volume of sales also relies on the technicians, owner, and manager. In fact, in some cases, a sale may result from this team effort. More specifically:

- The owner's visibility at a service facility can have a positive influence on customer confidence and, therefore, sales volume. Basically, the psychology of this logic is that although the automobile represents a large investment for the customer, the service facility is an even larger one for the owner. Because protecting his or her investment is important, the owner must be concerned enough about the customer's satisfaction to protect it. In addition, some customers prefer to speak to an owner just to be assured that he or she knows what is going on in the facility.
- Managers usually have the experience and awareness of the facility's "bigger picture." As a result, when managers oversee daily operations, they can offer different perspectives to a service consultant about a selling problem, give insights on the advertised maintenance specials, and step into difficult conversations to support the service consultant such as explaining a warranty or the financing details for a major repair. Over time, this influence supports sales and either increases or maintains the sales volume.
- The technicians, who first must be able and willing to properly diagnose and examine the customers' automobiles. Second, they must be able to communicate what they see to the service consultant in a way that the service consultant can pass the information on to the customer. If this information is not clear, technicians must be able to further explain the problem to the service consultant or, in some cases, directly to the customer. When technicians play a supporting role, the repairs that are hard to sell are often sold.
- The service consultants' attitude toward customers, their willingness to help them solve their repair problems, their skill in assisting them to make financial decisions, and their ability to build trust with them are critical to the sales volume. At the same time, however, they must be able to work with the members of the team. For example, they should

point out to customers how the owner is always "around," call in the manager to assist when customers need another perspective or a second opinion, and bring in a technician to confirm what the customer was told by the service consultant. The image being projected by the service facility is enhanced when customers know that several people are looking after their best interests.

An attitude of "going the extra mile" helps to build customer confidence and trust in service facility personnel. The personal connections and total concentration on the customers' best interests will, in many cases, build trust, credibility, and tomorrow's sales.

Review Questions

Multiple Choice

1. After inspecting a vehicle the technician recommends the following: replacement of a damaged driver-side seat belt, cooling system flush, replacement of brake pads that only have 1/32 inch remaining, and an oil change that is 1,500 miles overdue. Which of these represents the best way to prioritize this list to the customer?
   A. Brake pads, oil change, seat belt, cooling system service
   B. Seat belt, oil change, brake pads, cooling system service
   C. Seat belt, brake pads, oil change, cooling system service
   D. Oil change, cooling system service, brake pads, seat belt
2. If a customer objects to the cost of a necessary repair, which of these is the best response that the service consultant can give?
   A. Offer the customer a discount on the repair.
   B. Reschedule for a later time.
   C. Explain the reasons for the cost and benefits of the repair.
   D. Remove it from the repair order immediately.
3. Service consultant A says that providing a ballpark estimate is a useful tool to close a sale. Service consultant B says that being friendly and asking for an appointment will close a sale. Who is correct?
   A. A only
   B. B only
   C. Both A and B
   D. Neither A nor B
4. A 30,000-mile maintenance procedure is being performed. Which of these is a benefit of performing the service?
   A. The cooling system gets flushed.
   B. The transmission fluid gets changed.
   C. The maintenance will help the vehicle to continue to deliver dependable service.
   D. The completed checklist is given to the customer.

5. Service consultant A says that an example of a feature of an oil change is the weight and brand of oil used. Service consultant B says that an example of a benefit of an oil change is longer engine life. Who is correct?

   A. A only

   B. B only

   C. Both A and B

   D. Neither A nor B

6. An upset customer comes in when the service department is very busy. Which of these is the best way to handle the situation?

   A. Listen to the upset customer explain the whole problem.

   B. Ask the customer to come back later.

   C. Tell the customer he or she is wrong.

   D. Offer the customer a discount.

7. A customer has come to pick up his or her vehicle when the service department is very busy. Which of these is the best way to handle the situation?

   A. Direct the customer to the cashier to cash him or her out.

   B. Advise the customer that you are very busy.

   C. Review the work performed and the invoice with the customer.

   D. Ask the customer to come back when it is quieter.

8. When customers pick up their vehicle, Service consultant A says that it is important to take the time to explain the work performed in as much detail as the customer requires. Service consultant B says that if customers ask questions, it indicates they do not trust the shop/dealership. Who is correct?

   A. A only

   B. B only

   C. Both A and B

   D. Neither A nor B

Short Answer Questions

1. List the ways in which a service consultant can promote the procedures, benefits, and capabilities of the service facility and its employees.

2. What does "close the sale" mean?

3. Provide examples of service features and benefits.

4. Provide examples using features and benefits to sell "top-of-the-line" parts service.

5. List ways to overcome customer objections.

6. Explain how to identify and prioritize a customer's vehicle needs.

7. Explain how to present customers with the work to be performed and related charges.

8. Explain how different methods of payment can help close the sale.

PART II

CLINICAL PRACTICUM
EXERCISE

Visit a service facility and observe the service consultant both on the telephone and interacting with customers in person. How is the service consultant's presentation similar to or different from the information presented in Part 2 of the textbook?

Small Group Breakout Exercises

Your group owns and manages a chain of service facilities.

A. List what your group expects from a service consultant, and describe how you plan to make your customers feel at home so you can take care of their needs.

B. Create a service script, then practice it within your group. After it is perfected, demonstrate it to your classmates with a non-group-member classmate serving as the customer. If an in-class demonstration is not possible, record the interaction on audio or video. Listen to the interaction and compare it to your script. Repeat this process until you have an error-free recording or videotape.

Explore the World Wide Web

You are the owner of a service facility in a state of your choice. Search the Web for any information about consumer protection or consumer disclosure statements. (Hint: Look on your state's attorney general's Web site under consumer protection laws). Then summarize your findings and relate them to the automobile service industry and the service consultant's job.

PART III

INTERNAL COMMUNICATION, RELATIONS, AND SUPERVISION

CHAPTER 10

WRITING FOR THE TECHNICIAN

OBJECTIVES

Upon reading this chapter, you should be able to:

- *Demonstrate how to effectively communicate customer service concerns and requests to the technician (Task A.2.1).*

- *Demonstrate how to assist service facility personnel with vehicle identification information.*
  - ○ *Locate and utilize the vehicle identification number (VIN) (Task B.7.1).*
  - ○ *Locate the production date (Task B.7.2).*
  - ○ *Locate and utilize component identification data (Task B.7.3).*
  - ○ *Identify body styles (Task B.7.4).*
  - ○ *Locate paint and trim codes (Task B.7.5).*

Introduction

After a customer approves a service by signing the repair order, a copy of the repair order is given to the technician. This copy may be called the technician's worksheet, repair order hardcopy, or hardcopy. The popular term used for this copy, however, is *hardcopy*. This is because before computerized repair orders were popular, the last copy in a set of preprinted repair forms was made of light cardboard and was for the technician.

Today some computer forms print the technician's hardcopy on regular paper, while many shops choose to use the heartier cardboard. This is because it holds up in a shop environment, and the heartier material keeps it from getting bent and smudged from being around liquids and lubricants in the shop. The term hardcopy is used in this chapter to represent the technician's copy of the work order. To further protect the technician's copy, it should be placed on a clipboard. In many facilities, the clipboard is placed on a wall peg for the technician.

The front of the hardcopy is an exact duplicate of the repair order and must have all of the automobile and customer information needed by the technician. The purpose of this chapter is to cover the information that the service consultant must place on the technician's hardcopy. This includes the information that the service consultant wrote down about the customer's concern in terms that the technician can use to service the automobile. It also is to include information on the automobile.

Ensuring that all of the required information is entered on the repair order in the space where it is to be entered is important to the efficiency of the shop. Technicians cannot use their time to hunt for the information on the form, to interpret what is written, or to look for information on the automobile. This is because the technician is paid by the hour and there is an hourly charge for the use of the repair bay. Lost bay time costs the service facility. In addition, if the technician is paid on a flat-rate basis (explained in Chapter 3), this lost time costs the technician money and the service facility loses money. Service consultants must have their technicians in a position where they can immediately begin to work on a customer's automobile as soon as they receive the hardcopy.

Recording Customer Information: What and When

In some cases, the computer program provides the service consultant with space for a limited number of characters, such as 100. Therefore, what a customer has to say may have to be boiled down to 100 alphabetical letters, periods, and spaces. This limitation is preferred because it is inefficient for technicians to have to take the time to read a long report. To keep the description of the problem to the point, service consultants should focus on *what* the problem is and *when* it occurs. The service consultant's choice of words to state *what* and *when* is important because of the limited space provided.

For example, assume that Mr. Berger has a problem that needs to be diagnosed. The concern has not been verified and, therefore, the first words to record are: "Customer claims," which contains fifteen characters. The next part of the sentence needs to tell the technician *what* happens to Mr. Berger's automobile. In this case, assume that Mr. Berger states that his automobile will not properly shift from second to third gear.

The service consultant must ask Mr. Berger more questions to determine *when* the problem occurs. As the service consultant asks more questions, Mr. Berger explains that the automobile's engine speeds up before it goes into the higher gear. To put Mr. Berger's entire explanation of the problem into writing may exceed the number of characters the system will permit, plus the service consultant must select terminology used by technicians to describe the sensations felt by the customer. Specifically, the service consultant may use the word "slips" to describe *what* happens when the automobile's transmission shifts gears. To describe approximately *when* the problem occurs, the service consultant writes down when the problem occurs such as, "when the car shifts from 2nd gear." As a result, the description that is written on the repair order for the technician states: "Customer claims the transmission slips when the car shifts from 2nd gear." The statement is under 100 characters and conveys to the technician *what* the problem is and when it occurs.

The beginning of the sentence states "Customer claims" to indicate that the problem must be verified. As explained in Chapter 5, the verification of a problem is very important because under many warranty contracts, as well as state lemon law statutes, a problem does not exist until a technician verifies it.

Duplicating the Problem

After receiving the hardcopy, a technician makes some preliminary checks on Mr. Berger's automobile, such as checking the transmission fluid level and condition. Next he tries to duplicate the problem by taking the automobile for a drive but the problem does not reappear. Mr. Berger's automobile does not seem to have a problem when shifting from second to third gear. As a result, the technician must write down on the hardcopy that the problem was checked and could not be duplicated. The hardcopy is then returned to the service consultant, who presents the information to the customer.

When a problem cannot be confirmed, a number of actions may be taken. For example, the service consultant and the technician may take the automobile for a drive to see if the problem reoccurs. An option may be to ask Mr. Berger to drive the automobile with the technician riding as a passenger or vice versa. Another alternative that is often used when a problem is intermittent or occurs after the automobile has sat overnight is for the service consultant or technician to take the automobile for a drive after the lunch break or to take it home overnight. The objective in all cases is to duplicate the problem to obtain an accurate diagnosis.

Selecting Appropriate Terminology

Some service consultants may lack the technical knowledge or have trouble choosing the right words to describe a relatively detailed customer concern. In these cases, the service consultant should have access to a publication such as Delmar's *Automotive Dictionary* by South and Dwiggins (Thomson Delmar Learning, Clifton Park, NY, 1997). For service consultants at new automobile dealership service departments, the automobile manufacturer typically recommends terminology to use to describe certain concerns in the narrative, such as the transmission fails to engage upon acceleration, or service operation codes may be used.

Vehicle Identification Information

Service consultants must become very familiar with the location and formats of the VINs. The information contained in these numbers and letters is critical to the service consultant and technicians. For this reason, these numbers are to be included on all repair orders, which are copied on the hardcopies and invoices. The vehicle identification information is important to obtain correct parts; identify service procedures and specifications; and to obtain correct information about recalls, campaigns, and TSB information.

Other information to be recorded in the data file for the automobile and, therefore, on the work order is information on the automobile's optional equipment. Specifically, this information is needed so accurate estimates can be generated for customers. For example, when replacing an engine, the air-conditioning equipment adds time and cost to a repair. If the service consultant does not include this information in the estimate, the service facility, and possibly the technician, will lose time and money on a repair. Another example regards repairs performed under a warranty contract. These contracts require that the invoices include appropriate technical information, such as the automobile's transmission model, before payment is made.

Vehicle Identification Number

The VIN is a seventeen-digit number, and a special number is assigned to each automobile. The service consultant can locate the VIN at a number of different places. Once it is located, it should be recorded in the customer's file in the computer database.

The likeliest place to obtain the complete seventeen-digit VIN is from the automobile owner's registration card. The safest place to get the number, however, is to copy it from the place where it is located on the vehicle, which is expected in some warranty contracts. The VIN is most commonly found under the windshield near the driver's area but may also be located on the driver's door jamb.

Once the VIN is located, the service consultant must carefully copy the number down on a piece of paper. When copying the VIN, service consultants must be sure to write down each letter and number accurately. Failure to record all seventeen digits accurately will cause several problems, such as ordering wrong parts, rejection of warranty claims, inaccurate repair times, and even state law violations. One suggestion is that after the VIN is copied on a piece of paper, the service consultant should count the numbers to be sure there are seventeen.

What the VIN Digits Represent

An example of an automobile's VIN (1G2FS32P8RE100000) is shown in Figure 10-1. Note that this is a number used by General Motors. Manufacturers may use different letters and numbers in a category. For example, the digits shown in Figure 10-1 create the following eleven categories: origin (where the car was manufactured), manufacturer, make, vehicle line, body, restraint, engine, check digit, year, plant code, and sequence number.

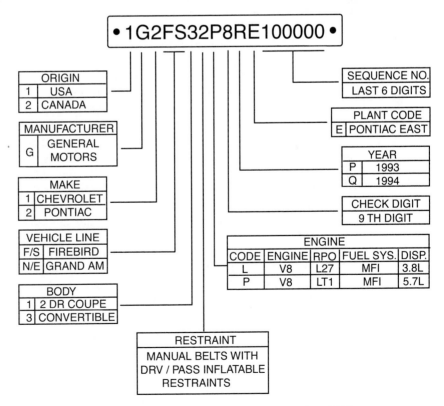

FIGURE 10-1 A vehicle identification number (VIN).

Figure 10-1 shows that the automobile was made in the USA (1) by General Motors (G). The vehicle make is a Pontiac (2) and the line or model is a Firebird (FS). The body style of the Firebird is a convertible (3) that has manual belts with air bags on the driver and passenger sides (2). The engine in this Firebird is a V8, 5.7 liter (P). The check digit for diagnosis is the 9th digit (8). The vehicle was manufactured in 1994 (R) at the Pontiac plant on the East Coast (E).

Again, caution is required when interpreting what a digit represents. For instance, the number for the body style in Figure 10-1 is a 3, which indicates a convertible. This number may not represent a convertible for other manufacturers because the numbers used to describe a particular body style may be different for different manufacturers. The numbers may be different even within a manufacturer. Also, the number used to describe a particular body style may change from year to year and may be different in a manufacturer's car lines and truck lines. Consequently, the meaning of each letter or number must be found in a shop manual or database.

Date of Production and Model Year

The production date of a vehicle is most commonly needed to order certain parts. For example, there may be two brake pad part numbers for the same automobile. The production date of the automobile is needed to decide which part number to use. The month and year an automobile was built can be found on the jamb of the driver's door. In addition, at new automobile dealership service departments, the production date along with a multitude of other important information can be obtained by entering the entire VIN into the manufacturer's database.

When ordering some parts, only the vehicle model year may be needed. At one time, automobiles made after September were assigned to the next model year. In other words, if a vehicle was made in November of 1995, it was considered a 1996 model. This is not the case anymore because the model year for an automobile changes when the manufacturer determines the "time is right." The timing for the change of model year is dictated by a number of factors that are associated with the manufacturing of the automobile. As a result, a manufacturer may decide that a new model year may begin at any point in the year, such as June.

To obtain the correct model year for an automobile, service consultants may look it up on the automobile owner's registration card, the 10th digit of the VIN, or under the hood of the automobile on the emission decal. When the emission decal is read, it will state that the automobile "meets the Environmental Protection Agency emission standards" for "model year," unless the hood has been replaced or the sticker is missing.

Parts Identification

Often service consultants and parts specialists need more specific component identification information to order a part. The reason is because even with the model year of the automobile, production date, and complete VIN, there may still be too many different parts from which to make a selection. Again, the dilemma the service consultant or parts specialist faces is that a mistake in a part order will cause delays, and customer expectations will not be met. In addition, the price for an incorrect part may cause a problem in the estimated repair cost.

Admittedly, in some cases, the quickest way to obtain the part information is from the old part itself. If a number cannot be found, an option is to remove the part and take it to the parts supplier for comparison to a new part or a picture in a parts book. If these methods fail, another option is to obtain the part number from a source that is specific to a manufacturer. For example, Ford Motor Company uses what it calls a calibration code to identify emissions parts. Ford puts the calibration code (see Figure 10-2) on a sticker located on the driver's side door jambs (late-model vehicles only). The calibration code then can be used to order the emissions parts for Ford Motor Company vehicles.

General Motors, in contrast, uses production option codes to describe the specific characteristics of each automobile produced (see Figure 10-3). The list of codes can be found in a production code manual, such as B72, which describes a type of molding used on 1996 to 1997 General Motors vehicles. In total, there are approximately 50 to 100 production codes for

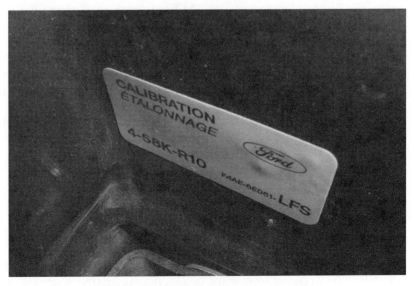

FIGURE 10-2 A Ford calibration code sticker.

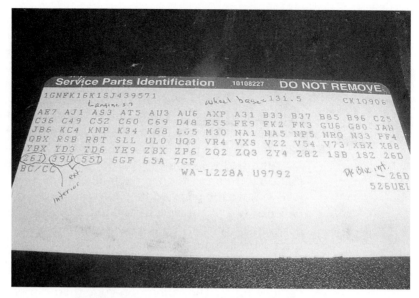

FIGURE 10-3 A General Motors option code sticker.

every General Motors automobile. The codes describe every characteristic of the automobile, from its mechanical parts to the trim and paint. The codes are so detailed and the list so complete on each vehicle that it is difficult to find two General Motors vehicles that have all of the same codes. Location of the GM Production Option tag varies from model to model. The most common areas to find the General Motors tags are the glove box or spare tire cover.

In addition, most manufacturers place the trim and paint codes on the driver's door jamb along with the VIN, the Gross Vehicle Weight Rating (the maximum weight of the vehicle after it is fully loaded with people and cargo), and other important information such as wheel base, tire size, and inflation pressures. Once the tag is located, the information or code needed to identify paint, for example, can be given to a parts supplier so the correct color can be obtained.

Summary

Working with customers to obtain correct information is one of the most important production tasks that a service consultant performs. The accuracy of the information and how clearly it is communicated to technicians and parts specialists is critical to a facility's efficiency. Becoming skillful in obtaining needed information is part of the service consultant's job.

In addition, service consultants must provide accurate information to all team members and be willing to work with technicians as partners in the repair process. The smooth processing of information and building effective working relationships among all employees is important to the quality and quantity of work performed by a facility.

In the communication process, service consultants must be able to locate information on automobiles as quickly as possible. They must know where to find the same information in at least two places. This is helpful if the information is missing at one location or if there is doubt as to the accuracy of the information.

Finally, a facility's workflow, which is covered in Chapter 11, is totally dependent on the accuracy and the speed at which the information is collected by the service consultant. If the information obtained by a service consultant is not correct or it is not on the hardcopy, workflow will be delayed and the time needed to complete a job will be extended.

Review Questions

Multiple Choice

1. A customer calls with a shopping list of problems with his or her vehicle. How does the service consultant put this information in a format that will help the technician find the customer's problem?

   A. Write down everything the customer says in the order he or she says it.

   B. Ask open-ended questions regarding each item to determine the problem.

   C. Ask the customer to boil the problem down to a specific system on the car.

   D. Verify that each item on the repair order is a symptom or a maintenance request.

2. Which of these is a common location for the production date?

   A. On the valve cover

   B. Inside the driver door pillar

   C. On a sticker on the radiator support

   D. Inside the gas door

3. Service consultant A says that before computerized repair orders were popular, the last copy in a set of pre-printed repair forms was made of light cardboard and was for the technician. Service consultant B says the word "hardcopy" often is used to describe the technician's copy of the work order. Who is correct?

   A. A only

   B. B only

   C. Both A and B

   D. Neither A nor B

4. Service consultant A says that when replacing an engine, the air-conditioning equipment will NEVER add more time or cost to a repair. Consultant B says when information such as air conditioning is not considered in an estimate, the service facility, and possibly the technician, will lose time and money on a repair. Who is correct?
   A. A only
   B. B only
   C. Both A and B
   D. Neither A nor B

Short Answer Questions

1. Explain how a service consultant can effectively communicate customer service concerns and requests to the technician.
2. List ways a service consultant can assist service facility personnel with vehicle identification information.
3. Go to a vehicle, then locate and record the following:
   A. the VIN
   B. the production date
   C. various component identification data
   D. the trim package (GT, LS, etc.)
   E. the paint and interior colors

CHAPTER 11

WORKFLOW

OBJECTIVES

Upon reading this chapter, you should be able to:

- *Explain the relationship between company policy, operational procedures, and operation manuals.*

- *Define workflow.*

- *Present a plan to manage customer appointments. (D.5)*

- *Describe techniques to use to manage workflow. (D.1)*

- *Explain why monitoring workflow is important. (A.2.5)*

- *Prepare a progress chart for monitoring repair orders.*

- *Describe the reasons behind active delivery and the use of a formatted system.*

Introduction

Workflow is the processing of work from the initial contact with the customers to the return of their automobile. The ultimate objective for an automobile service facility is for all work to flow smoothly through the facility. To accomplish this, service consultants must: 1) control the volume of work taken into the facility each day, 2) monitor and facilitate the flow of work through the facility, and 3) have the customer's paper work and automobile prepared for pickup as soon as work is completed.

For service consultants to control and facilitate the flow of work, they must have a firm understanding of the facility's process and procedures. To discuss these processes and procedures, the chapter begins by describing the creation of **policy** and how it is used to write **operational procedures and regulations**. Next, the chapter explains how and why operational procedures and regulations become the guide for the generation of **operation manuals** that are to be used by employees to process work through a facility.

The chapter then builds upon the service consultant's duties presented in the earlier chapters to discuss the workflow process, such as the procedures service consultants follow to make appointments and prepare work orders. In order to meet the objective to have all work flow smoothly through the facility, the service consultant's duties must be linked together into a seamless operation. The seamless operation is then at the heart of the operations manual, which is like a rule book.

In fact, some franchise service facilities have carefully developed operation manuals to the point where they require each independently owned franchise to use them. The benefit is the assurance that their customers are provided with exactly the same services or products each time they come into one of the facilities, regardless of the location. Because of the success of some franchises, their operational procedures have set the standard for their industry. For example, Ray Kroc founded McDonald's and worked hard to make sure customers knew and got what they expected at each McDonald's franchise around the world. Their procedures made McDonald's a leader in the fast food industry. As a result, the concepts of exact procedures that make up a system were copied by other industries and businesses. The procedures that make up a system begin with the company's policies.

Company Policy

The owner, owners, or a corporate board of directors set **company policies**. These policies indicate how the owners want to conduct business, although some of them come from a law or legal regulations, such as the consumer protection laws and state safety inspection regulations. A policy statement may explain why the owners want the policy to be followed, but this is not necessary, because owners and boards are not required to justify their directives to employees.

Policies should also describe the authority and responsibilities of the managers and service consultants. These descriptions should be in reference to the positions of the managers and service consultants in the organizational structure, which was discussed in Chapter 1. As a consequence, when service consultants begin a new job, they should carefully review their responsibilities, which are usually shown in a job description. In fact, the service consultant should review the job descriptions of all employees with whom they work. These descriptions are typically approved by the owner(s) or board and are critical to the workflow process.

As shown in Figure 11-2, company policies lead to the operational procedures and regulations of the service facility. The operational procedures

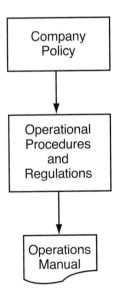

FIGURE 11-2 The connection between policy and operations manual.

and regulations, which usually are written by the owner(s) or managers in charge of operations, are the basis for the operation manuals (the rule books) to be followed by all employees. The operations manuals direct how the work is to be processed by each work area. Therefore, to ensure that the owner(s) or board expectations as well as the law will be met, there is a direct link between the operation manuals and company policies. As shown in Figure 11-2, there may be one manual for the entire facility with sections for each work area or a separate manual for each work area in the facility, such as the parts department. This chapter will assume that there is a manual for each work area.

An example of a policy that goes into an operation manual is as follows.

1. A company policy states that: *All parts put on a customer's automobile must be purchased through the facility.*
2. Operational procedures and regulations may then direct that: *All parts needed for customer automobiles must be ordered by the parts specialist.*
3. The operation manual may then direct that:
   i. *All parts will be ordered by the parts specialist within one half hour of the order being placed by a technician.*
   ii. Employees understand the full sense of the policy that: *Customers may not bring their own parts to the service facility for installation.*

Some directives in an operation manual for a work area will overlap with other work areas. In the example above, the operation manuals for the service consultant, the technicians, and the parts department specialist would all state that: "Customers may not bring their own parts to the service facility for installation."

Beginning the Workflow Process: Scheduling Customer Appointments

The workflow process begins with the scheduling of customers for work. If there are too many appointments, some customers may not have their work completed by the end of the day. If there are too few appointments, some technicians may not earn their expected wages for the day and service facility sales will be lower. To further complicate the process, if a service facility has some technicians who perform maintenance work and some who perform repairs, the service consultant must actually schedule work for two operations. Therefore, even though the customers should always be encouraged to make an appointment, service consultants must estimate the time needed for each service so the customers' needs are met.

To illustrate how this works, first, when scheduling an appointment, the service consultant might say, "I can schedule you to come in at either

8:15 or 9:45 on Friday. Will either of those times work for you?" Experienced service consultants know that, if possible, a customer should be given two times to choose from because it encourages them to make a decision and it gives them a choice. Service consultants also know to schedule customer appointments at least 10 to 15 minutes apart, or in some cases farther apart when more in-depth problems must be discussed. Otherwise, customers will be impatiently lined up at the service consultant's workstation.

Service consultants should never tell customers to "just stop in when the shop opens at 8 a.m." or, worse yet, tell them, "first come first served". While some service facilities, such as the fast lube industry and many tire retailers, can operate without appointments, service facilities that offer a range of services must have greater control. If the service consultant does not have some control over when customers come to the shop, they are prone to bunch up at one time and become impatient because they must wait in line.

Making Appointments

To schedule customers for appointments, a computer or an appointment book is used to keep a daily customer log. The advantage of a computerized appointment schedule is the ability to look at the customer's history in the database. As a result, when an appointment is being scheduled, suggestions for any additional services needed for the automobile can be made. Also a computerized appointment schedule can be updated from one day to the next and appointments can easily be changed.

When an appointment book is used, each sheet in the appointment book represents a new day, and there should be enough sheets in the book to cover the current as well as the next month. Also, since an appointment book requires the service consultant to write down the information, a pencil should be used so that changes can be erased. Pencils must be sharp and the handwriting must be neat so that others can read the entries. If a consultant's handwriting is not easily read, then all entries must be printed.

Estimating Repair Hours Available for Sale

Whether a computerized system or appointment book is used, the service consultant must record the estimated number of labor hours needed to perform the service for each repair order. For example, assume that Renrag Auto Repair had 3 repair technicians with approximately 20 flatrate repair hours of capacity to sell each day. The total estimated hours on the repair orders, therefore, should not exceed 20 hours.

The number of flatrate hours of capacity for a facility is based on the past performances of the technicians plus a few hours for diagnosis work and unexpected emergency visits. As a result, the capacity, or average number of hours a service facility may produce a day, is important for a service consultant to know in order to schedule the correct amount of work each day. For example, if Monday had a carryover of 4 hours of unfinished work to Tuesday, the capacity at Renrag Automobile Repair for Tuesday would no longer be 20 hours but 16 hours. Of course, the amount of time the service consultant should budget will vary with the type of work done at the service facility as well as the time of the year. For example, when the weather gets very cold or very hot, more time may need to be budgeted for unexpected repairs.

After any carry-over work, diagnostic work, and unexpected repairs are considered, a service consultant may have 12 out of 20 hours left over for appointments. In addition, a prudent service consultant will not want to schedule appointments that will take up all 12 hours because some of the appointments will require more work than recorded in the appointment book. To illustrate this point, let us say an average of 1 hour of extra time is needed for every customer who schedules an appointment. Assume David Paterno calls in to have his oil changed and tires rotated. In this case, the service consultant knows the labor time associated with this maintenance will take $1/2$ hour. This means the service consultant should budget $1/2$ hour for the service and 1 hour for any additional work David's automobile will need. As a consequence, after David's appointment is made, there are 10.5 hours available for more appointments. If each customer was similar to David, the service consultant could schedule a total of 8 appointments until the daily work capacity is filled (see Figure 11-3)

While the example in Figure 11-3 is easy to follow, the process is not quite as simple as it seems because additional time may be needed for an automobile. For example, an older automobile is likely to require more time than an automobile that is new. In such a case, the automobile's history will help the service consultant to assess the condition of the automobile and allow additional time. In addition, the history can help the

---

SUMMARY OF RENRAG AUTOMOBILE REPAIR'S TUESDAY'S SCHEDULE

Total capacity of labor hours at Renrag: 20 flat-rate hours

Holdover work from Monday	4 hours
Customer unexpected emergencies	4 hours
8 scheduled customers ($1/2$ hour each)	4 hours
Additional work (8 scheduled customers @ 1 hr. each)	8 hours
TOTAL flat-rate hours budgeted for services:	20 hours

Unbudgeted hours leftover = 0 hours

FIGURE 11-3 Calculation of daily work capacity.

service consultant identify additional repairs the automobile might need. For example, if the consultant saw in the computer database that the automobile had low brake lining readings on front and rear brakes during the last visit, the automobile will likely need new brakes. This can add another 2 to 3 hours of additional work to the schedule for this one automobile. Also, if the customer is new, then time should be allotted for an inspection (as described in Chapter 4) as soon as the automobile enters the shop.

The Daily Customer Log

The above example given for Renrag Auto Repair demonstrates that a service consultant cannot fill up every available hour of the entire workday with appointments. Otherwise there will not be any time left for work left over from the day before, diagnostic work, unexpected customers with immediate concerns and requests, and recommendations of the technician for work found during the service of scheduled customers. Keeping an accurate daily customer log is an important tool for the service consultant.

The minimum information needed in a **daily customer log** or appointment book is determined by the nature of the work done by the service facility. In general, however, the daily log should show the date, time of appointment, customer's name, phone number, vehicle, description of the work needed, and time of pickup (see Figure 11-4). Then the service consultant should estimate how long the service will take and add it to the time taken by the previous appointment (the last column shown in Figure 11-4). In other words, as shown in Figure 11.4, Paterno's work will take 1.5 hours and the estimated amount of time for Ricardo's work is 3.0 hours. So the total time scheduled for the day is 4.5 hours.

Date: _____

Time	Customer Name	Phone Number	Year & Make of Vehicle	Services	Pick Up	Est. Work Hours
8:30	Paterno	237-3333	2002 Pontiac	Oil change Tire rotation	Noon	1.5
8:45	Ricardo	655-4441	2001 Buick	Brakes	3:00	4.5

FIGURE 11-4 A daily customer log.

Finally, once the appointment book is filled, appointments for the following workdays should be scheduled. In this case, a service consultant might tell a customer, "I am sorry, but our schedule for today is filled; can I make an appointment for you for tomorrow at 8:30 or 8:45a.m.?" If the customer indicates that the next day is not possible, the service consultant should suggest two later dates and times from which the customer might choose. In other words, care must be taken to not "over schedule" the facility and create unnecessary pressure and dissatisfied customers whose delivery expectations will not be met. At the same time, to "under schedule" work for the facility will result in lost profit. Therefore, scheduling work in advance should be encouraged whenever possible.

Workflow: Service Consultant's Paperwork

When a customer's automobile is ready to receive a service, the service consultant must start the paperwork introduced in Chapter 4 and then discussed in the following chapters. An operation manual needs to direct what paperwork is to be filled out and how it is to be conducted for each step of the flow process shown in Figure 11-5. For example, the operation manual would direct in step number one how the customers are to be entered into the database.

The operation manual would also direct when a diagnosis is to be performed and when the hardcopies of a repair order are to be distributed to the technician. As shown in Figure 11-5, a request for a diagnosis (if needed) is sent to the technician before an estimate is prepared for the customer. The operation manual would indicate who should receive the request from the service consultant.

Workflow: The Technicians

While an operation manual for technicians may include specific procedures about how to perform certain repetitive jobs, such as diagnostic procedures for no-start, no-crank problems or maintenance services such as oil changes, it must also state how work is to be processed after it is received from the service consultant. An example of a flowchart showing how a repair order is to be processed is shown in Figure 11-6.

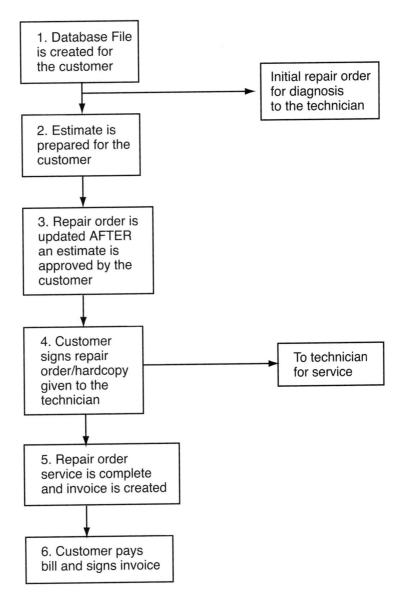

FIGURE 11-5 The flow of paperwork at a service facility.

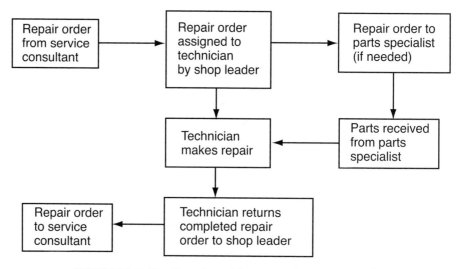

FIGURE 11-6 The flow of work for processing a repair order.

In the flowchart in Figure 11-6 there are several important features to be noted since they would not be the same at all service facilities.

1. There is a shop leader who receives the repair order from the service consultant.
2. The shop leader assigns the work.
3. The technician leader places the order for parts with the parts specialist.
4. When the parts are received, they are delivered directly to the technician.
5. When the work is completed, the technician returns the repair order to the shop leader (who, in turn, assigns another job).
6. The shop leader returns the completed work order to the service consultant.

By following a flow diagram as shown in Figure 11-6, a written set of procedures for an operation manual could easily be prepared. For example, the operation manual would start with the service consultant generating a repair order, which goes to the shop leader, who assigns the work to technicians. The operation manual should also describe how technicians obtain automobiles for repair, check them over, write down their comments about the repairs, return repair orders to the shop leader, and pull automobiles out of the shop; where to park them; and how to get another repair order.

Monitoring Repairs

The operation manual should require the service consultant and/or the shop leader to keep a log on the status of all repair orders. Only by keeping track of the progress of repair orders can service consultants monitor their facility's workflow.

To effectively keep track of the many repair orders in a given day, the service consultant should use a progress chart. Progress charts can either be a part of a computer program matrix or be hand recorded on a form in a notebook or on a clipboard. The purpose of a progress chart is for the service consultant to be able to follow the progress of multiple work orders at the same time. As a repair order is transferred from one employee to the next, the status of the order is recorded on the progress chart, which may also be called a **repair order tracking sheet**. An example of a progress chart (called a repair order tracking sheet) is shown in Figure 11-7.

To use the repair order tracking sheet shown in Figure 11-7, a service consultant must create the repair order. The service consultant will then record the customer's name, the automobile make and/or model (vehicle type column), the repair order number assigned to the job, the completion time promised, and the description of the work on the sheet. This information is necessary because it is common for the service consultant to think of the job and refer to it in terms of the customer's name. However, technicians will often think of the job and refer to it in terms of the make and model of the automobile. Furthermore, parts specialists will often think of the job in terms of the types of parts ordered for the repair or

Repair Order Tracking Sheet

Customer Name	Vehicle Type	Repair Order Number	Time Promised	Work Description	Parts Clerk Estimate	Customer Approval	Parts Arrived	Work Finished	Invoice Prepared	Customer Notified	Vehicle Picked Up

FIGURE 11-7 A progress chart on repair orders.

even a repair order number that must be recorded on parts delivery receipts. The use of different information to describe the same job can cause miscommunication, and the repair order tracking sheet can act like a score card to connect the different references to a job.

As per the workflow sequences described earlier in this chapter, the repair order would be given to the shop leader to assign to a technician. Once the technician checked the automobile and recorded his findings on the repair order, it would be returned to the service consultant, who would record the time the repair order was received and forwarded to the parts specialist for the estimate on the repair order tracking sheet.

Next, the parts specialist would give the service consultant the parts and labor estimate. The service consultant, after meeting with the technician, would call the customer. Each time the service consultant tries to call the customer, the time of the call should be entered on the repair order. After the service consultant gets approval for the job, the time of the approval should be entered in the customer approval column. The service consultant must also note on the repair order the time of the approval, the person who gave the approval, the phone number called, what was said, and what was approved (depending on state law, other information may also be required).

Next, the service consultant returns the repair order to the shop leader to order the parts. The parts specialist must notify the service consultant when the parts arrive in order to record the time in the "parts arrived" column. The parts specialist then returns the repair order to the shop leader, who reassigns it to a technician (typically the same technician will receive the job), and the parts are delivered to the technician. Once the job is assigned to a technician, the service consultant should be notified so the time can be recorded in the "work being finished" column. After the work on the automobile is completed, the repair order should be returned to the service consultant so an invoice can be created. As a further note, the procedure should require the technician to use a time clock to punch on the repair order when work is started and finished. After an invoice is completed, the time should be written in the "invoice-completed" column along with the time when the customer was notified the automobile was finished. Finally, the time "active delivery" occurred is written in the "vehicle picked up" column.

The time recordings on the tracking sheet are important to the monitoring of repairs by the service consultant. For example, this information is helpful when estimates are needed on the amount of time until a customer's automobile is ready for pickup, if an emergency repair can be taken into the facility, and if there might be time for a diagnosis. In addition, if the service consultant suspects a problem exists with a repair because it has taken longer than the estimated time or if the parts have not arrived as expected, the customer can be called and advised of a possible delay.

By monitoring the tracking sheet, service consultants can be aware of changes that might cause a job to be carried over to the next day or that an additional job can be accepted. For example, assume that at the beginning of the day, the service consultant has scheduled all of the daily work possible. However, at noon the service consultant finds that two jobs scheduled for the day are still waiting for parts to arrive. When a customer stops in for a maintenance service, the service consultant may be able to bring in the job that day as opposed to scheduling it for another day because of the parts delay.

Note that the times should be recorded on the repair order when customers are called for approvals. For example, assume the tracking sheet indicates a job begun at 8:30 a.m. did not get an approval from the customer until 11:30. In the case of a complaint about the delay in making a repair, the service consultant may examine the repair order to determine when the customer was first called. If calls were made to the customer at 9, 9:30 and 10:15 a.m., then obtaining approval was a problem and a new strategy may be needed to reach customers. If the first phone call to the customer was at 11 a.m., however, then the customer was not contacted as directed in the operation manual.

The shop leader as well as the parts specialists may also use a tracking sheet. They may modify the sheet to meet the particular needs of their work area, such as distributing work to technicians and tracking the order, receipt, and distribution of parts. In all cases, the operation manual should direct the procedures for each work area.

Active Delivery

The end of the workflow process occurs when customers are presented with an invoice and the keys to their car. This delivery is an important function that is often overlooked by service facilities. The service consultant must remember that for most customers, the automobile is the second most expensive item they own (a home is the most expensive item) and they depend on it to take care of themselves and their family.

Pre and Post-Inspection of Customer Automobiles

Of course, before a service consultant makes an active delivery, the customer's automobile must be checked. This requires service consultants to have the pre-inspection form (shown in Chapter 4) on which the technician notes the general condition of the automobile, including damages and missing or broken parts. Consultants should then inspect the automobiles and compare them to the pre-inspection report before delivery to the customer. Next, consultants should ensure that all repairs and services were performed, the vehicle was cleaned, no tools or old parts were left in it, and it was detailed (assuming this is a customer policy).

Therefore, the care and cleanliness of the customer's automobile should be monitored throughout the service process. This requires the use of fender covers, seat covers, floor mats, and steering wheel covers. Needless to say, upon delivery of the automobile to the customer, the automobile should be at least as clean as when it arrived.

Shop rags and old parts should not be left in the automobile (or under the hood) and all protective covers should be removed for the customer's convenience. A customer is not impressed when the paper floor mats are left in the car because there is no place to put them. Customers are even less impressed when there are oil stains on the carpet and not on the paper floor mat. In addition, the automobile should not have any grease stains on anything that a technician touched, such as white wall tires or door handles.

The Delivery

The primary goal behind an active delivery is not so much an action (such as washing of the automobile) but an attitude taken by the service consultant. The service consultant should take the time to add a personal touch to the delivery process.

Active delivery can include driving the automobile to customers or escorting them to their automobiles, which should be in front of the facility's entrance. The service consultant should express appreciation for their business, provide them with the opportunity to check over their automobiles, and allow them to state concerns or requests they may have. This then permits the service consultant to correct any misconceptions or problems the customer may have as well as to point to the positive points about the customer's automobile. Then the service consultant should ask if the customer would like to schedule any additional work.

Summary

When operation manuals present the workflow procedures to be followed by each work area, the expectation is that all employees will perform each of their job tasks correctly. This is important because: 1) each job task is a piece in the workflow process, 2) all of the job tasks must fit together in order to create a process, and 3) in order for the tasks to work together to create a smooth workflow, the employees must work with each other as a team. When this occurs, work can be conducted both effectively (the job gets done correctly) and efficiently (little waste occurs, so a profit is earned).

The preparation of the operations manual is admittedly time consuming to create; however, automobile service facilities that clearly state their operational procedures and regulations in a manual (or rule book) to employees are more likely to be successful. One reason is because customers know they can count on the service to be received every time they come to the facility. Some refer to the dependence on a manual as a **"formatted system"** because it promotes consistency throughout the automobile service facility. The recognition of consistency, in turn, gives customers confidence that if they were properly served previously, they will be properly served again.

Review Questions

Multiple Choice

1. A customer has just given approval for repair of a vehicle. Consultant A says the technician should be provided with the approved work order. Consultant B says documentation of the customer's approval should be on the work order. Who is correct?
   A. A only
   B. B only
   C. Both A and B
   D. Neither A nor B

2. During quiet times when customer and phone demands are low, a service consultant should:
   A. Examine the progress of customers' car repairs and try to anticipate any problems that might occur.
   B. Get a head start by examining what aspect of their job can be done before the next wave of customers arrives.
   C. Take a break and either sit in the manager's office or visit the technicians.
   D. A or B is correct.

3. A customer approved the diagnosis of a problem but has denied approval for repair of their vehicle. Service consultant A says the technician should not perform the repair. Service consultant B says documentation of the customer's denial to have the repair performed should be written down on the work order. Who is right?
   A. A only
   B. B only
   C. Both A and B
   D. Neither A nor B

4. The workflow process begins with the scheduling of customers for work. Service consultant A says if there are too many appointments, some customers may not have their work completed by the end of the day. Service consultant B says if there are too few appointments, some technicians may not earn their expected wages for the day. Who is correct?

   A. A only

   B. B only

   C. Both A and B

   D. Neither A nor B

5. When booking an oil change appointment for a good customer, the consultant finds that his highly paid drivability technician is the only one with openings on that day the customer wants. Which of these is the best solution to this problem?

   A. Book the job for the drivability technician.

   B. Add it to the lube tech's schedule and tell the customer you will work them in.

   C. Offer the closest time that does not conflict.

   D. Move the appointment of a new customer.

Short Answer Questions

1. Explain the relationship between company policy, operational procedures, and operation manuals.
2. What is workflow and why is it important to the service consultant?
3. How would a plan to manage customer appointments be different at a small shop versus a large shop?
4. List techniques a service consultant could use to manage workflow.
5. Why does a consultant need to monitor workflow?
6. How is a progress chart used to monitor repair orders?
7. What are the benefits of active delivery?
8. How can a formatted system at a service facility help the productivity of the facility?

CHAPTER 12

CUSTOMER RELATIONS: SALES, FOLLOW-UPS, AND PROMOTIONS

OBJECTIVES

Upon reading this chapter, you should be able to:

- *Demonstrate how to greet customers and respond to angry customers (A.1.6).*

- *Explain the importance of identifying and prioritizing customer concerns (A.1.3, C.2).*

- *Give examples of how to promote the procedures, benefits, and capabilities of the service facility (A.1.8).*

- *Identify methods to communicate the value of performing related and additional services (C.4).*

- *Describe methods used for customer follow-ups (A.1.5).*

Introduction

The previous chapters cover in detail the processes and procedures to be followed when working with customers. The purpose of this chapter is to focus on customer relations. Specifically, this chapter discusses basic sales techniques and follow-ups, methods used to promote sales, and handling angry customers.

Selling Services

As explained in previous chapters, service consultants have several opportunities to promote sales when they come in contact with customers. These include the initial contact on the telephone, when the customer comes into the shop, after the technician's initial inspections and diagnosis, when the automobile is serviced and problems are noticed by the technician, and during active delivery. During these contacts, the job of the service consultant is to sell what the service facility offers. In time, service consultants develop their own style of selling; however, there are several basic principles to be recognized.

The first principle is to always offer customers a pleasant and friendly greeting. When a person calls on the phone, the caller must be treated with courtesy and the conversation should be given proper attention. The service consultant must speak clearly and distinctly.

To ensure that phone conversations are handled properly, the phone should not be located where there is a lot of background noise. This can cause communication problems for both the caller and the service consultant. If cordless phones are used, the service consultant should not attempt to carry on conversations in the shop area where technicians are working on automobiles.

When customers enter the facility, service consultants must offer an appropriate greeting as follows:

- "Welcome to Renrag Auto Repair."
- "My name is _____, and your name is?" (if possible, extend the right hand to offer a firm and friendly handshake while waiting for the customer's name)
- "How may I help you?"

After a proper welcome, customers will state what they need, which could include the repair of a problem, maintenance, or just general information. When customers state their concerns, service consultants must be good listeners and give them their full attention. Service consultants must listen carefully to the customers' problems, what the customers think, and what has been done (or what the customers think has been done) to their automobiles. Service consultants must ask "when," how often," and "where" questions. Most importantly, sales cannot be made if service consultants are not aware of what the customers need, and, more importantly, wish to buy!

Selling and Angry Customers

A critical sales technique for service consultants to develop is how to handle angry customers (a more in-depth discussion about dealing with angry customers is included later in the chapter). First, the service consultant must always treat the customer, including angry ones, with courtesy. Second, service consultants must keep in mind that the objective is to close and complete a sale. This interaction should be handled like a business meeting! Third, the service consultant must realize that when a customer is angry, it is typically about the inconvenience and unknown costs (anticipating the worst) associated with a repair. It is not personal.

Service consultants must realize that customers rely on their automobiles to meet their personal needs. When their automobile cannot be used, even for a brief period of time, customers and their families will experience transportation problems and will not be able to take care of their daily obligations as easily. This may cause feelings of helplessness, agitation, and frustration that cause some people to be angry. Because these encounters are not common, service consultants must learn to interact in a disciplined and positive business manner.

Customer Personalities and Sales

After service consultants learn what a customer's problems and concerns are with respect to their automobile, they should prioritize the work as discussed in Chapter 9. In addition, they should focus on the features of the service and benefits received as also discussed in Chapter 9. Service consultants must provide customers with accurate, detailed explanations of the service to be purchased and answer all questions. The focus should be on the capabilities of the facility to take care of the customer's needs.

When discussing a service with a customer, service consultants should attempt to adapt their discussion to the customer's personality. For example, if a person seems to be anxious, cynical, and/or distrustful, the consultant should allow plenty of opportunity for the person to ask questions, give thoughtful answers, and permit the customer to have more time than usual to make a decision (in other words, the customer should not be pressured to give an answer). If a customer is not interested, is in a hurry, and/or seems bored with the description of the service, the service consultant should provide explanations that are more limited. If a person is irritated and/or upset about the problem, the service consultant should use the business approach that revolves around the features of the product, benefits of the service, and repair priorities.

Overselling and Up-selling

There is often confusion about when up-selling becomes overselling. Overselling means a customer is sold a service that is not needed. For example, a manufacturer's manual may recommend a maintenance service and part replacement at 50,000 miles. If a new part is not needed and the automobile has traveled 25,000 miles since the last service, the customer has been oversold.

Up-selling is when a customer needs a maintenance or repair service and has an option on the quality of the parts or the amount of labor put into a repair. For example, assume that a customer has to have new brakes. The choice is between less expensive, lower-quality brake pads and more expensive, higher-quality pads. The customer would obviously save money by buying the cheaper pads; however, the person would have to pay to have them replaced sooner. The cheaper pads may also have different warranty coverage. If the higher-quality pads are purchased, the customer would save money in the future and may have better warranty coverage, and the facility would make more profit on the sale of the better set of pads. In the long run, the customer will be more satisfied with the repair. Hence, the service consultant should try to up-sell the customer on the purchase of the higher-quality brake pads.

Up-selling also occurs when a customer's automobile is having a maintenance or repair service and the technician sees a maintenance or repair that should (not must) be performed. For example, assume that an automobile is having an oil change and the technician sees that the windshield wipers are so worn that they will not pass a safety inspection. If the service consultant sells new wipers, it would be an up-sell. If the wipers are not worn out, however, it would be an oversell. Oversells are how facilities get a bad reputation, eventually losing customers, and they may break a state's consumer protection law!

Customer Disagreements with Sales Proposals

If customers do not agree with a recommended repair or set of repairs, the service consultant should ask for his or her recommendation. In some cases, the customer may want to purchase more than what the service consultant is trying to sell! If the service consultant does not agree with a customer's suggested repair because it will not solve the problem and the person continues to insist, the service consultant should suggest that the person seek another opinion.

In one situation, a customer wanted organic brake pads when his car required semimetallic pads. This was a problem because organic pads do not last as long or bring the automobile to a stop as well as the semimetallic pads. The service consultant refused the customer's request. The customer then offered to buy the pads and bring them to the facility for installation. The service consultant again refused and the customer left. This proved to be a good decision because after the customer had the pads installed, he rear-ended another automobile when his did not stop as quickly as it did before the replacement.

Refusing to make an improper repair at the request of a customer might result in a lost sale, but it is preferable to risking a comeback, a breakdown on the road, an accident, or a lawsuit. Any of these results would cost a facility more money than it would make on the repair.

Promoting the Facility

Service consultants must constantly promote the facility and the personnel who work there. They must have a positive attitude (we can fix your car) and must be the company cheerleader (we are the best). In other words, they must always be selling the facility and the technicians' ability to fix automobiles. They must tell customers about the good reputation of the facility, customer compliments, how their technicians go to extra lengths to solve complicated problems, the quality of the parts put on customer automobiles, and the benefits of allowing the facility to keep customer automobiles in top shape. Service consultants must also promote the capabilities of the facility, the expertise of the technicians, and the use of expensive equipment to maintain and repair automobiles.

In addition, if the technicians and service consultant are ASE Certified, the display must be placed on the front window, on a building sign, in newsletters and specials, on professional name cards, on uniforms, and in all advertisements. At the same time, the names and positions of the employees who are ASE certified must be prominently displayed where the service consultant greets the customers. At Renrag Auto Repair, the ASE displays impressed customers, many of whom commented or asked questions about ASE.

Another display that attracts attention is a bulletin board. This board should have copies of all specials, advertisements, and newsletters. Furthermore, it provides a place to post thank-you cards, publicity, photos, items of information, and results from customer surveys. When customers are waiting for their automobiles, most take the time to look at this board; the service consultant should encourage those who do not.

Sales Follow-Ups

The service consultant's job is not over after the customer's automobile has been serviced and the bill has been paid. Rather, another phase of the job has just begun. This is the "follow-up" part of the job and is closely related to the promotion of sales. Service consultants must be concerned about the customer's satisfaction if they wish to sell them services in the future.

The primary objective of the follow-up, however, is to identify as soon as possible those customers who are NOT satisfied with their service or treatment. If a customer is unhappy about a service, the likelihood of making a future sale to that customer is not good. It is therefore important to find out why the customer is dissatisfied and to correct the problem if possible, and then to avoid creating the same problem in the future. Furthermore, if a dissatisfied customer tells other people about the bad experience, serious damage to the reputation of the facility is likely to occur.

When service consultants realize that a customer is not satisfied, they must collect as much information as possible about the complaint. They must check if other customers have the same complaint and then investigate the matter. If there are any weaknesses, errors being made, or employees not doing their job properly, then corrective action must be taken as soon as possible. After actions are taken to correct a problem, the service consultant should call the customer, or customers, who made the complaint to inform them of the corrections, express appreciation for their interest in assisting the facility, and invite them to return for service (possibly with a discount).

First Follow-Up

Customer follow-ups begin with the "thank you" during the active delivery of the customer's automobile. At this time, the service consultant might say:

- "Thank you for allowing us to work on your automobile. If there is anything that was not satisfactory about your experience or service, would you please let me know?"
- (Pause to permit the customer to respond.)
- Then continue with, "If you would be willing to permit us to service your vehicle in the future, please call me."
- (Then hand the customer a business card and possibly a self-addressed, prepaid postage customer satisfaction survey card, described next.)

Second Follow-Up

After a week or so, a follow-up "thank you" is recommended. This form of appreciation may be: (1) sending a thank-you card to the customer; (2) mailing the customer a satisfaction survey form with a thank you card; or (3)making a personal phone call to ask about the automobile's performance. In addition, regular reminders should be sent to customers inviting them to schedule an appointment for a future service, especially seasonal specials.

Several companies sell attractive thank-you and service reminder cards to send to customers. At Renrag Auto Repair, because these cards were quite popular, they were changed from season to season. Most importantly, however, many customers expressed their appreciation for the attention when they returned to the facility.

The Personal Touch

Another follow-up involves personal recognitions. These personal touches help build an ongoing relationship with customers so that they feel quite comfortable when they come to the facility. They also help to make conversation, but they require service consultants to remember customers.

One suggestion is for service consultants to keep a notebook on their customers. The notebook contains the name of the customer with notes about conversations, names of children, special interests, and so on. For example, when customers return to a facility, a service consultant may mention how much he or she enjoyed a story told at the last visit, ask them about their children by name, or wish them luck in a bowling tournament or fishing trip.

Sales Promotions

The primary objectives of **sales promotions** are to keep regular customers and attract new ones. Different types of sales promotions are discussed in the next sections. Some will be more successful for a facility than others. In addition, two important points to be recognized when promoting sales are (1) a method may not show any results until it has been used several times and (2) each promotion must be designed to fit the facility; for example, a specialty shop has a different type of customer than a general repair facility.

Newsletters

Newsletters are sent to customers. They can contain articles of interest (Figure 12-1) and useful information about automobiles, as well as reminders and specials the service facility is offering to customers. The purpose of newsletters is to develop a relationship with customers so they will return to the facility for service.

Newsletters are a benefit because some customers may forget where they had their automobile serviced last. Newsletters, therefore, can serve as nice reminders. Studies have shown that it is less expensive to get a previous customer to return to a service facility than to attract a new one. As noted earlier in the book, regular customers keep a facility in business.

How Did Mandatory
State Vehicle Inspection Start?

by Dr. Ronald A. Garner

While automatic transmissions and air conditioning were introduced because of consumer demand, some new systems were introduced as a result of government regulations. The primary reason for government regulations was to ensure public health and safety. One of the first government concerns was the high number of vehicle accidents during the 1940s. In 1944, one study conducted in Cincinnati (population 500,000) found that there were 5,201 accidents and 151 were caused by faulty automobile equipment such as lights, brakes, or steering. At the same time in the State of California, 2,703 people were killed in automobile accidents, which was more than those killed at Pearl Harbor (*Automotive Digest*, 1945). These findings caused some state governments to begin to require periodic automobile inspections. For example, in Pennsylvania, mechanics must complete a training program and then pass a written and hands-on exam before the state issues a license to perform safety inspections.

FIGURE 12-1 A newsletter article of interest.

Service Reminders

In addition to newsletters, facilities should send out service reminders to inform customers of maintenance services due on their automobile (see Figure 12-2). The creation of a reminder is a fairly simple task that can be done using computer systems that have a database with up-to-date customer addresses and service information. Some of these systems can automatically print monthly reminders for customers whose safety inspection is due. In some cases, a discount might accompany the reminder to further entice the customer to return for a service.

Before sending out service reminders, service consultants should review their content as well as the names of the people to whom they will be sent. For example, a customer whose check bounced (meaning the account did not have enough money in it to pay the bill) should not be sent a reminder with a discount. Also when reviewing the names, the service consultant can handwrite personal notes on some of them.

SERVICE REMINDER

...that will ring your bell

Dear Valued Customer,

Our records show that your car is due for its **annual Pennsylvania state safety and emission inspection** in the upcoming month. We understand how much you depend on your car and how difficult it is to fit service into your busy schedule. But if you call us now, we will help to make this state mandated inspection service as convenient as possible. So, call 555-5555 NOW to make an appointment:

Schedule before the 10th of the month
and as a special "thank you"
we will take $6 off any other regularly priced service
(oil change, coolant flush, alignment)*

*If performed at the same time. Applies only to regularly priced "menu" services,
not valid with any other specials or discounts.

If you have a friend or relative who also needs this service, tell them to call us and we will gladly extend the special "thank you" discount to them, as a courtesy to you. Again, thank you for being such a good customer and we hope to see you soon.

THANK YOU
Put in the owner's, manager's, service writer's, or technician's name

FIGURE 12-2 An example of a service reminder.

Customer Specials

Customer specials are sent to regular as well as potential customers. **Specials** should be attractive, have a catchy heading, and serve a particular purpose. For example, a special may be offered because of a change in weather (see Figure 12-3) or because of a large inventory purchased at a reduced price, such as a large, bulk delivery of oil. A special may be offered on a particular day of the week, week of the month, or month of the year when business is slow. Another special may promote a new product line or service.

Hot Summer Special

Dear Valued Customer,

Because (**insert Business Name**) values good customers like you, we wish to say "thank you" by giving you a Summer Special. Summer weather will soon be upon us and we want to help you and your car stay cool. As a result, we are offering you this special. CALL (**555-5555**) NOW to make an appointment before (**MONTH 1, 200X**) and we will perform our:

HOT SUMMER SPECIAL

- Check ALL of your A/C system's parts
- Check A/C system for leaks
- Check refrigerant level and add up to one pound
- Perform a # point summer inspection of your car
- Inspect your car's cooling system
- Give you advice about problems your car may have

Normally $ _____ NOW $ ____ **for R-134 systems**
Normally $ _____ NOW $ ____ **for R-12 systems**

Most Cars and Light Trucks
Referigerant dye extra if required

If you know someone who also needs this special, tell them to call us at (555-5555 by Month 1, 200X) and we will let them have this summer special at this price as a courtesy to you. Again, thank you for being such a good customer and we hope to see you soon.

THANK YOU,
(Put in the owner's, manager's, service consultant's or technician's name)

FIGURE 12-3 An announcement of a special.

Because specials usually have a low profit margin, they should be promoted in the most inexpensive way possible. One method is to print the information on colorful paper to create flyers that are handed out, placed on windshields, sent through bulk third-class mail, placed in high traffic areas, enclosed in newspapers, and so on. The objective is to gain high visibility at the cheapest price.

Another method to advertise specials is to place them on large advertising signs in front of the facility. These signs may be purchased at department and parts supply stores and are not expensive. They can be lighted and be maintained by employees of the facility. The special announcements, of course, must be short and to the point; for example, "Oil Change, Lube & Filter—$19.95."

A more recent means to promote specials is through e-mail. Many businesses ask customers for their e-mail addresses to send them specials and reminders. This is the cheapest means to promote specials and maintain contact with customers. The facility can use its computer system to prepare and send out the announcements.

Public Advertising in Newspapers, Radio, Television, and Solicitations

As many service consultants soon learn, advertising sales representatives are regular visitors at a service facility. They want to sell advertisements in newspapers, radio stations, television, restaurant paper placemats, church bulletins, public school and college sports programs, high school yearbooks, special charitable events, car racing, billboards in stadiums, and so on. Advertising is an absolute necessity for a business; however, it is expensive, complicated, and may not be effective unless done properly.

Because of the many types of public advertising methods, a facility should consider employing an advertising consulting firm. First, it can put a balanced advertising plan together. A good consulting firm also has or can obtain the number of people and their demographics (e.g., age, gender, etc.) that an advertising source reaches. Second, an advertising firm can screen all advertising sales requests made to a service facility. This removes the service consultant and managers from having to deal with advertising representatives during business hours. It also keeps them from choosing one business over another and possibly losing a customer. In these cases, the representatives are simply referred to the advertising firm. And third, the consultants can prepare and review all public advertisements to ensure proper content and appearance. For example, the size of a newspaper ad and the words, pictures, diagrams, and placement in the paper are critical to the success of the promotion.

Without an advertising consultant, an automobile service facility must make all of the decisions about where, when, and how to advertise. Public advertising must be seen by potential customers and fit the needs of the facility. For example, an independent repair facility may be more interested in advertising to people who own older, as opposed to new, automobiles that are under warranty. In addition, there is no need to advertise if a facility is operating at 85–100+ percent of capacity for a particular service or product.

When public advertising is being considered to attract new customers, then the design of the type, mix, and amount of advertising must be carefully prepared. Each type has its advantages and disadvantages. If service consultants are responsible for public advertising, they must carefully study the types of advertising available to the facility and then prepare an advertising campaign.

Customer Survey

Customer feedback is an important assessment tool often used to determine if customers are pleased with the services they received and the work performed on their automobile. As noted previously, some feedback should be gained during the active delivery process; however, many facilities use a formal customer survey. For example, some service facilities:

- mail the survey, which is on a prepaid postage card, to the customer. (This method is preferred by many automobile manufacturers.)
- leave the survey in the customer's car to be mailed back to the service facility or other location. (This method is commonly used by organizations representing the facility, such as AAA.)
- call the customer on the phone to ask survey questions. (This is often used by larger dealerships or companies that perform market research.)
- give the survey form to the customer to fill out and place in a locked box before leaving the facility. (This is often used by independent service facilities and dealership service departments.)

Survey designs and the method used to obtain useful information may be complicated; however, most have a number of common features. In general, the questions in a survey look at three basic questions: (1) whether the repair bill was more or less than what was expected; (2) whether the automobile was finished on time; and (3) whether the services met expectations.

The survey questions usually ask customers to rate the facility on a scale of 1 to 5 (1 being the least satisfied and 5 the most satisfied). The survey may also ask a series of yes and no questions in order to look for a pattern of responses. For example, if the responses on a scale (positive) did not agree with the yes/no answers (not positive), a problem is likely to exist and the customer deserves some attention such as a follow-up call to extract more information.

Analysis of Survey Information

After the surveys are collected, a customer service index score can be calculated. This index consists of the average score for each question, the average of all of the questions, the range of answers for each question, the number of choices for each number in the scale, and the number of yes and no answers. The index is then compared to previous surveys for the facility and to those of other service facilities.

In some cases, advanced statistics are used by a market research company to identify patterns. For example, automobile manufacturers can compare their dealer's scores with each other and with competitors' service departments. In some cases, questions and their answers may reflect on a specific employee or group of employees. In some instances, this score is so important that service consultants may ask customers who do not give them an excellent rating what they can do to meet their expectations and industry standards for excellence. Therefore, larger companies (such as automobile manufacturers) track customer surveys over a period of time to determine if any changes in perceptions of a facility occur and how they compare to those of the facility's competitors.

To get accurate information, some companies collect multiple surveys from the same customers to determine if a change in marketing emphasis or service procedures is warranted. For example, a company may survey customers shortly after visiting the service facility for the first time, then again several weeks or months later. The purpose is to see if the customers' perceptions have changed after they had more time to think about their responses.

While customer satisfaction ratings are an important assessment tool for a service facility, it is not necessarily the magic bullet for success. There are many variables that go into customer service to produce superior satisfaction ratings as well as higher profits. The idea that "doing anything the customer wants" or offering gimmicks typically erodes profits that must be paid through higher labor rates and markups on parts. In these cases, a service facility may find that it cannot compete in the marketplace. Therefore, caution is necessary when a person believes there are no limits on how far a company should go to improve customer satisfaction.

Dealing with Angry Customers

Some customers express their anger at the time they pick up their automobile, whereas others pick up their car and drive it home and then unleash their anger the next day. Their anger is typically because they believe the service facility did not meet their expectation. More specifically, the customer may believe that the automobile was not serviced properly, that the invoice was too high, that the service facility forgot a request that was important to him or her, or that the automobile was not returned in the same condition as when it was left at the facility.

First, the service consultant must determine if what the customer believes is true. When a customer is rightfully angry, the service consultant should immediately apologize. The service consultant should follow up with explanations (but not excuses) and/or immediate action to determine why the problem occurred and what can be done to correct it. One action may be to bring the automobile back to the service facility to examine why the service did not meet the customer's expectation. The problem may be as serious as a part that was not installed correctly and has to be reinstalled, or it may be as simple as a grease spot that must be removed.

The service consultant should consider the following steps to get through an inherently bad situation when a customer is clearly angry:

1. Welcome and greet the customer in the same manner as a person who is not showing signs of anger.
2. Allow the customer to vent his or her frustrations; therefore, do not make an attempt to interrupt.
3. Do not take anything the customer says personally. Remember, the problem is frustrating the customer, not you.
4. As the customer begins to exit the stage wherein his or her frustrations have been vented, examine the situation and assess the real problem in a professional business manner (do not get sidelined into unrelated issues).
   A. If a customer becomes abusive, the service consultant must stop the interaction and ask the customer to refrain from using abusive language.
   B. If a customer refuses or does not seem to be able to exit the frustration phase so the issues at hand can be addressed, the customer must be asked to leave.
   C. If a customer does not leave, possibly because of being under the influence of drugs or alcohol, local law enforcement should be called to handle the situation before it escalates out of control.
5. Focus statements and the conversation on the real problem.
   A. Start this process by summarizing why the customer is angry.
   B. Ask the customer to agree with your assessment of the problem so that he or she knows you understand.
6. If the problem was in some way caused by the service facility, offer an apology immediately. For example, if a stranded customer needed a tow truck and it arrived late because the service consultant forgot to call the towing company after the customer called, an apology is necessary.

7. Understand company policies and how far a service consultant can go to help the customer.
   A. Watch what is said and how it is said. Do not make statements that are outside the authority of the service consultant's position.
8. Provide a solution or several options to rectify the problem within the authority of the service consultant's position.
   A. Most service facilities give service consultants a variety of tools to help in a situation. At some facilities, service consultants can offer a "free or almost free" warranty repair under certain circumstances and/or a **policy check**, which has "make believe" money in it, for credit toward a future purchase (see Figure 12-4). For example, a repair may be just beyond the mileage requirement called for in a warranty, yet circumstances indicate the warranty should be honored, so a policy check may be given to the customer. A policy check would not be given or a warranty service would not be honored, however, if a customer caused the damage, such as filling a gasoline tank with diesel fuel.
   B. Ask the customer to suggest a solution or choose an option suggested.

If a customer does not accept any of the solutions or suggestions offered to correct the problem or requires financial compensation for the trouble and anguish caused, then the problem must go to the next level of authority. When management has to be involved, service consultants must document everything that occurred, and the conversation with the customer must be reconstructed as accurately as possible. This documentation will be critical if the complaint should escalate to the point that legal actions will be taken by either the service facility or the customer.

In some cases, a complaint may not be the fault of the service facility and would not require an apology. This sometimes occurs because customers refuse the services recommended. For example, a customer at Renrag Auto Repair had an antifreeze leak that was coming from a loose bypass hose at the front of her engine. The hose was relatively new and the clamp simply needed to be tightened to fix the problem. However, upon closer inspection, it was found that her heater hoses were very old (original by all appearances) and worn in some spots. The technician recommended replacing the hoses but the customer refused the suggestion. She did not feel that it would be a problem even though her car was over 10 years old and had over 120,000 miles on it. A couple of weeks later, her heater hoses blew apart on the freeway. The customer continued to drive her automobile until the engine seized. The angry customer demanded a new engine from Renrag Auto. After a lengthy discussion with her and her attorney, which required an extensive review of the notes on her invoice and all conversations, she agreed that the service facility was not in the wrong. Although unfortunate, the problem occurred because she did not take the recommendation of the service consultant at the time of her service.

When an invoice charge is greater than expected (possibly justified) and disputed, service consultants may be able to use a policy check with the "make believe" money in it. This permits a service consultant to adjust the amount of the invoice by writing out a policy check (see Figure 12-4) for the amount of the disagreement. The amount on the policy

Policy Check

Renrag Auto Repair
(Address)
555-555-5555

No: _____

Date: _____

Pay to the

order of _____

$ _____ Dollars

**Good only for credit toward a repair performed at Renrag Auto Repair.
Void without proper validation and authorized signature. (No cash value)**

Authorized Signature — RAR

Reason _____

R/O number _____ Date _____

FIGURE 12-4 A voucher for a policy check.

check is either deducted from the invoice in dispute or the customer presents it for payment toward a future service.

In some cases, a policy check may not be effective. For example, a customer needed an oil change and a wash because she was going to a funeral the next morning. The service facility made the oil change and found a recall that needed to be done. The recall did not pose an immediate threat and could have been scheduled at a future date. The service consultant forgot, or disregarded, the customer's request and told the technician to perform the recall. The service facility finished the oil change and the recall by closing but ran out of time to wash the automobile. The customer was angry because her car was still dirty. Neither a policy check, nor a free oil change, nor a review of the services performed would make this up to the customer. Therefore, an apology was all that could be offered. The service consultant had to recognize that listening to the requests of the customer is the most important part of the sale.

Summary

Sales depend on having customers. Creating a good, solid customer base is one of the largest costs to a new business. Attracting customers costs a lot of money in advertising and promotions. Also, while a facility is waiting for customers to come in for service, the employees do not have work. Income is not being generated and other expenses, such as wages and electricity, are being incurred. As a result, many new business owners purchase an existing business with regular customers as opposed to starting a new business from scratch.

The point is that (as has been stressed throughout the book) the recording of customers into a database is critical to a facility's future sales. Furthermore, the maintenance of the customer database must be given the attention it deserves. For example, the addresses, phone numbers, and e-mail addresses must be up to date. When a newsletter, announcement, or e-mail is returned due to a change in address, the customer must be removed from the database. People who are not good customers (don't pay their bill) must also be removed from the database. Printing and sending out mail is expensive and time consuming. Retaining poor customers damages the work environment (see Chapter 14), and the ability to stay in business.

Given a service consultant has many tasks to perform and makes promises to customers and employees throughout the day, organization is key. To forget a promise or call a customer at a time told at the least will cause disappointment and at the most anger. To help avoid this, service consultants should keep a small notebook where they can write down what they need to do and promises made. As each is done, the service consultant should cross it off his list. At the beginning of each day, a new list should be started so promises outside of the normal routine are remembered.

Review Questions

Multiple Choice

1. Which of the following is considered an up-sell?
   A. Higher-quality brake pads that will last longer
   B. Replacing all fuses in a fuse box with ones that are not blown
   C. Removing air from the tires and replacing it with new air
   D. Installing a new battery because the old one is bad

2. Two service consultants are discussing customer follow-up. Service consultant A says that their purpose is to identify customers who are NOT satisfied with their service or treatment. Service consultant B says it is important to find out why the customer is dissatisfied, correct the problem if possible, and then avoid the same problem in the future. Who is correct?
   A. A only
   B. B only
   C. Both A and B
   D. Neither A nor B

3. The primary objective(s) of sales promotions is/are to:
   A. attract new customers
   B. keep regular customers
   C. Both A and B
   D. Neither A nor B

4. Which of the following is NOT a method to obtain customer feedback?
   A. Mail the survey to the customer.
   B. Leave the survey in the customer's automobile (hung on the mirror) to be filled out and sent in later.
   C. Hand the customer the survey when he or she pays the bill.
   D. Have the technician call the customer.

5. When discussing a service with a customer, Service consultant A says that the discussion should be adapted to fit the customer's personality. Service consultant B says that the customer should always be pressured to give an answer. Who is correct?
   A. A only
   B. B only
   C. Both A and B
   D. Neither A nor B

Short Answer Questions

1. Explain how to greet customers and respond to those who are angry.
2. Explain why a consultant must identify and prioritize customer concerns.

3. Give examples of how to promote the procedures, benefits, and capabilities of the service facility.
4. Identify methods to communicate the value of performing related and additional services.
5. Describe the different methods used for customer follow-ups.

PART III

CLINICAL PRACTICUM EXERCISE

Visit a service facility and observe the service consultant. How does he or she manage time and the pressures of his or her job? How is the promotion of sales performed and how is it similar to or different from the information you learned in Part III?

Written Communication Exercise

Using a computer develop a one-page customer mail-out that promotes your service facility. Use the information below to build a special to offer your customers. Add some graphics to make it appealing to your customers.

Information to include in your special: "You are a good customer and we want to say thank you by offering you this special. This winter will be cold and we want to make sure you and your car stay warm by checking your car's cooling system and heater. CALL XXX-XXXX to make an appointment and we will:

- Check ALL of your hoses and cooling system's parts
- Check the cooling system for leaks
- Check antifreeze level and condition
- PLUS perform a 10-point winter inspection of your car
- Provide "advice about your car"

Small Group Breakout Exercises

List what you must personally accomplish over the next three days on a piece of paper. Consider the following in your "to-do" list.

- Where must you be?
- What must you study and read?
- What projects must you work on and how much of it must be done (what goal must you reach)?
- What errands must you complete?

After listing the information, use a new piece of paper and organize the activities listed from the most important (must get done) to the least important. Discuss in your group why you organized each activity as you did.

As you complete each activity over the next three days, cross it off. After three days, meet back in your group to discuss what you completed on your list and how you would reprioritize the past three days relative to what you got done.

PART IV

COMMUNICATION: CUSTOMER DELIVERY AND FOLLOW-UP

CHAPTER 13

OTHER DUTIES: GENERAL OPERATIONS ASSIGNMENTS

OBJECTIVES

Upon reading this chapter, you should be able to:

- *Outline the procedures for making arrangements for a sublet sale (D.4).*

- *Describe the various "other duties" a service consultant may have to perform.*

- *Explain the purpose of a petty cash account.*

- *Explain the reason for liability insurance and workmen's compensation.*

- *Describe the security systems that can be implemented at a facility.*

Introduction

Up to this chapter, the text has pointed out the primary responsibilities and expectations for service consultants. Those who have worked as a service consultant, however, can testify that there are a number of "other duties" assigned to them. Some of these duties are formal job assignments, while others are assumed because of the physical location of the service consultant or his or her relationship to other employees. In other words, because service consultants are located where they will meet people as they come through the door, they get most of the questions, requests, problems, and, sometimes, even deliveries. At times, they can refer the person to someone else, but they often must take care of the person's concern.

In addition, other employees, such as the technicians and parts specialists, often go to the service consultant when they have a need or problem——sometimes for a problem as simple as a new lightbulb to replace one that has burned out. The reason is because the service consultant can easily be contacted and knows where "things" are kept. Thus, it is natural for employees to go to the service consultant to take care of general problems. For this reason, the service consultant may be referred to as the "go to" person.

In many cases these "other duties" make the "go to" service consultant indispensable. Employees and customers know that top-notch service consultants get things done, and this can make them important. Unfortunately, it is impossible to list all of the tasks or duties that service consultants may perform. In addition, these "other duties" vary from one facility to another. For example, although the actual tasks may be the same, the ones performed by a service consultant at a small independent repair facility are likely to be different than those performed by a service consultant at a large dealership or a franchise. The purpose of this chapter is to introduce the various general business-related duties that may be assigned to service consultants at different types of service facilities.

Business Contracts

One set of "other duties" assigned to service consultants and described next is the result of contracts between the service facility and other businesses. Service consultants rarely have the authority to enter into any business contracts on behalf of the facility; in fact, for their own personal protection, they should not enter into any business contracts for a facility unless given a direct order to do so by management or the owners, preferably in writing.

Even though service consultants do not make contracts with other businesses, they must know their contents. This is because they must legally work within their agreements. They must know what the contracts state they must do, what the other businesses will do, how it must be done, and when.

Sublet Sales

The first agreements with other businesses to be discussed are arrangements for **sublet sales.** This is because service facilities are often quite dependent on the maintenance of good relations with a business to whom they sublet sales as well as with a business that sends them sublet sales.

There are many types of sublet sales arrangements. A facility may make arrangements through a written contract or a gentleman's agreement to do business with each other as long as everyone is satisfied. For example, assume that a service facility does not have a tow truck. Rather than buying one and being called out at all hours of the night, the facility makes an agreement to have a towing service take care of all of its customers who need assistance. When it receives a call from a customer who needs to have an automobile towed to the shop, the service consultant calls the towing service. The tow truck driver picks up the automobile and delivers it to the service facility. The service facility will charge the customer for towing the automobile and then pay the towing service business the fee. In many, if not all, sublet sales, the service facility that makes the arrangement for the service adds a fee to the charge. For instance, if the towing service charges $35 for towing the automobile, the service facility may add $35 to the fee and charge the customer $70 on the invoice.

Under a "gentleman's agreement," if one of the two businesses is dissatisfied with the arrangement, they simply end it. For example, assume a towing service decided to double its charges to the service facility. The service facility could simply make arrangements with another towing service. If the two businesses had a legal contract to do business with each other, the charges for towing would be in the contract and could not be increased until the contract expired.

Sublet sales arrangements may be made for a variety of services, such as muffler repair, radiator repair, sound systems, alignments, tires, body repairs, computer diagnostics, and others. Often customers are not aware that a repair was made as a sublet sale unless a warranty agreement was a part of the sale. In most of these cases, however, the warranty is provided to the service facility that arranged the sublet sale. So if a warranty is provided, customers do not go to the business to which the sublet sale was given (and made the repair) but rather to the service facility where they took their automobile. The service facility would then go to the business that made the repair to have the warranted work done.

Maintenance Contracts and Repairs

Maintenance of the building and equipment is an ongoing challenge that cannot be avoided. For example, maintenance contracts for the service of some equipment, such as the update of the alignment rack, must be done at a time when it will not disrupt business. The service consultant, of course, is probably the only person aware of the time when this service can be performed. When a piece of shop equipment breaks down, such as the air compressor, the service consultant must be directly involved because it can cause delays and appointments may have to be cancelled.

When equipment and facility maintenance and repair service personnel come to a facility, they often go to the service consultant for instructions. The service consultant must try to ensure that the maintenance and repair work is done as quickly as possible and does not disturb management and shop employees. When the work has been completed, the service consultant must often verify that the job was completed. When the service consultant is the contact person, he or she must understand the maintenance agreements and contracts. Examples of contracts/agreements at a facility include the communications systems, garbage collection, recycling of waste materials, security system, TV cable for the reception area, magazines, heating and air conditioning, computers, shop equipment, and lawn and landscaping care.

Uniforms and Cleaning of Uniforms

Uniforms for technicians and service consultants may be purchased or rented by the facility from another business. When they are rented (and in some cases purchased), they are picked up and cleaned once a week. The arrangements are usually clear-cut and may go through the service consultant. As a result, when problems occur, the service consultant may become involved; for example, if a technician does not have the right pair of pants delivered, the service consultant is often the one who has to "keep an eye out" for the delivery truck. Likewise, if the cleaner is short a pair of pants, the delivery person must go to the consultant to ask the technician where the pants are located. Therefore, the service consultant usually has to be aware of the agreements made for the ordering and cleaning of uniforms.

The presentation of the employees to the public at a service facility is important. At some service facilities, the service consultants are provided with a uniform or shirt with the name of the facility, the name of the employee, and, sometimes, the employee's job title. While this may not seem important, helping customers recognize an employee's position can avoid confusion. For the service consultant, being supplied with uniforms and laundry services can be a financial advantage, especially because they will likely get oil or grease on them from handling parts.

Suppliers of Parts and Goods for Resale

Service facilities must make arrangements with other businesses from which they purchase parts and a variety of goods, such as oil and tires, to sell to customers. These arrangements are usually made by the managers, owners, or business representatives of the company and typically require an approval for credit purchases. In return, the suppliers typically sell their goods to a service facility at wholesale, as opposed to retail, prices.

The suppliers of parts and goods to a facility are referred to as **vendors**. When a state levies a sales tax, the tax is not charged for parts and goods to be resold by another business. Rather, the sales tax is collected when the service facility charges the customers for the parts and services sold to them.

A service consultant must understand the business relationship and policies the facility has with different vendors. For example, consultants must know the return policy of the vendor and how old parts removed during service, called **cores**, should be returned to the vendor. In addition, since the inventory of parts kept at the facility (such as oil filters) may be important for production, the service consultant must know about the resupplying arrangements made with each vendor. While a parts specialist may order the parts, in some cases, the inventory is restocked automatically. If the restocking of the parts is not done by the time the service consultant makes a sale, problems will occur. With respect to parts and supplies at a franchise service facility, service consultants must be thoroughly informed of the arrangements set up for the purchase of parts and goods by the franchise corporation.

Service consultants must also learn how to obtain parts from new automobile dealerships. This is because the auto parts stores often do not have all of the parts required to fix newer automobiles. As a result, a service facility must have arrangements with the different automobile manufacturer dealerships for new automobile parts. It is not always common for dealerships to deliver parts so the service facility must be able to pick them up. For example, if a new automobile is being repaired, the service consultant must order the part (that is hopefully in stock at the dealership), and then make arrangements to pick it up. This sometimes affects the estimated time needed for a repair.

Petty Cash

Another responsibility that may involve service consultants is the **petty cash fund**. Although service consultants do not usually manage the petty cash fund, they have to know how it works and when to use it. Petty cash is kept on hand to pay for items that must be purchased from a vendor that does not have a credit arrangement with the service facility. For example, a service facility may need to purchase a fluorescent light from the hardware store. The employee making the purchase can obtain the cash needed from petty cash, make the purchase, and return the receipt to petty cash. Another option is for the employee to pay for the light and then submit the receipt for reimbursement.

Petty cash accounts are not to be used for major purchases because the amount of cash placed in the account is usually $50 or less. The money should be kept in a locked drawer or special locked box and only one employee and a manager or owner should have access to it. The drawer or box should be replenished periodically, such as at the end of each month, or when the cash available gets below a set amount, such as $10. When replenished, the amount of cash left in the drawer or box, plus the total receipts, must equal the amount placed in the account, such as $50.

Petty cash can be a nuisance but is invaluable when a small purchase is needed or a delivery must be paid for in cash. For instance, when a delivery is made to the service consultant and the person making the delivery needs to be paid for postage or a fee before the package can be left at the facility, cash is needed. Without a petty cash account, the service consultant will either have to personally pay for the delivery or not accept it. Either option is not desirable.

Liability Insurance and Workers' Compensation

Although liability insurance and workers' compensation are not a responsibility assigned to service consultants, they must understand what they are and how they work. **Liability insurance** covers the customers' automobiles while at the facility (called **"garage keepers insurance"**), a customer who may be accidentally hurt while at the facility, and other possible damages caused by accidents or negligence.

If a customer or his or her automobile is harmed while at the facility, service consultants must be aware of the procedures the insurance company expects to be followed. In addition, some customers ask about the coverage of their automobile when they leave it at a service facility. Service consultants must be able to explain the coverage, assuming the facility has a policy. For example, in one case, a technician was test-driving a customer's new automobile and accidentally ran into a dump truck. The roof of the vehicle was torn off and the car was considered demolished. The insurance company paid for the replacement of the customer's vehicle.

With respect to other accidents or negligence, damages may occur to a vehicle that does not belong to a customer, or a person may be hurt crossing the property of the facility. For example, a technician got into a customer's automobile and, forgetting the brakes were not working, drove it into a 9-foot wire fence. The impact knocked a fence post into an automobile sitting on the other side. The insurance company paid for the damage to the fence and both cars.

Workman's compensation is an insurance policy that covers the costs of injuries received by the employees of the facility while performing their jobs. This cost may include the pay lost by the employees due to the injuries. For example, an employee injured his eye while making a brake repair. The insurance company paid the emergency room medical bill. The company paid the employee for the half day of lost wages due to the injury. In this case, the employee was injured because he did not have safety glasses on while making the repair. After the accident report was filed, the insurance company instructed the owners to require all employees to wear safety glasses when making repairs. In addition, safety slogans were provided by the insurance company, which also required that they be posted throughout the facility. After several weeks passed, the insurance company representative checked to see if their requests were enforced.

Of course, if a business has an insurance claim, it is often due to negligence. When a claim is filed with an insurance company, it is investigated. When an accident occurs and someone or some property is injured, the insurance premiums may be adjusted. Safety slogans and hazardous warning signs, such as "Customers may not enter the shop area," should be posted. Failure to post the signs and heed insurance company directives can be costly to a facility.

Utilities

Water and electricity are essential to an automotive service facility. Usually a service consultant does not have any need to be involved with these services until a problem arises. In this case, it is important to know where the water shutoff valves and electric circuit breakers are located. When a utility service is disrupted or any building repairs are made, the people sent to make a repair must be shown where the control valves and panels are located. Again, because the service consultant is usually the first person people meet inside a facility, he or she is usually asked for assistance.

Banking

When service consultants receive payments from customers, they should know how to make deposits to the bank account. There should be two rules regarding the handling of cash at a service facility. First, there should be a limit as to the amount of cash kept in the register. Once the amount (say $1,000) is exceeded, a certain amount (say $200) should be deposited to leave an acceptable balance. In the case of a large company, the money may be placed in a vault. Second, at the end of the business day, only the amount of cash needed for opening the next day (say $200) should be left in the register overnight. This usually requires a night deposit into a bank's drop box.

To follow these two rules, service consultants must fill out bank account deposit forms and a daily earnings report, and then make deposits to the facility's account. If a manager or cashier is employed by a facility, he or she usually handles the reports and deposits. However, because cross training is key so that employees can cover for each other, the service consultant is often the backup for the cashier.

Advertising and Promotions

When advertising and promotions are prepared, service consultants are often asked for input. For example, do they think customers will buy a combination of specials at a certain price? In addition, when an advertisement or a promotion is being prepared, they may also be asked for their opinion on the presentation of the information, such as the layout and the words being used. A service consultant may even be asked to check the original before it goes to print. Finally, after the advertisement or promotion is released, phone calls from customers asking about information in the announcement will be directed to the service consultant.

Since service consultants answer inquiries about the services offered by a facility, their opinion and observations are logically sought out. So to make these promotions successful, service consultants should always make notes about what customers want, like, and dislike. Furthermore, to successfully answer a customer's questions about a particular advertisement, they should always have a copy of the ad handy.

General Supplies

Service consultants must keep an inventory of office and other supplies on hand and may be required to order them from a supply store. For example, they must be sure to have extra appointment books on hand, estimate-repair forms for the computer printer, printer cartridges for the computer, pens (sometimes with chains so they do not "walk away"), paper, credit card tapes (when they collect money from customers), phone message pads, tags to put on automobiles waiting to be serviced, key tags, and so on. In addition, at some smaller independent and franchise facilities, the service consultant may be expected to monitor the first aid kit as well as supplies needed in the restrooms.

In some cases, service consultants are responsible for maintaining the customer waiting area. This may include making coffee and even, in some cases, having pastries on hand in the morning. When the coffee detail is assigned to the service consultants, they must make sure the proper supplies are on hand and that the room is neat and clean.

In other words, one of the first things service consultants must do after being hired is to determine what supplies they must monitor and how they order them. Next, they must check the inventory to see how much is available. Finally, they must bring their inventory supply up to an appropriate level. They must monitor supplies as they are used. Running out of some supplies can be terribly disruptive.

Security

Most automotive repair shops have some form of a security system. At Renrag Auto Repair, which was located in a small city, there were two security systems. One was on the doors and windows and was monitored by a private security firm. The other one was a set of video monitors at the service consultant's work area connected to cameras in the service bays. The service consultants had to know how to turn on and off the door alarms and the TV system on and off when opening and closing the facility. They, as well as the other employees, also were expected to keep an eye on the monitors to identify when nonauthorized people were in a service bay.

Security is important to a service facility for several reasons. One is that they have money, expensive tools, equipment, and supplies to protect. Protection is needed when the facility is locked up at night and during the day when it is open. Automotive facilities are particularly vulnerable to theft because the large doors to the service bays are often open. In addition, a number of emergency exits must be located throughout the building. On hot days these emergency doors may be opened for ventilation. At some facilities, a plastic yellow chain is placed across open doors to keep unauthorized visitors out of restricted areas.

Another reason for a good security system is to reduce insurance costs. For example, customers are not permitted in service bays. Insurance companies usually require signs stating that customers are not permitted in the work areas to be placed on all bay doors. In addition, to protect the insured equipment and tools, they like to have alarms placed on the doors and windows. Furthermore, they are quite pleased when the bays are being protected through the use of a TV security system. In some cases, a good security system can reduce insurance premiums.

TV monitors for the service bays generate customer interest. At Renrag Auto Repair, the TV security monitors were located at the service consultant's workstation. Because the service bays could not be seen from the waiting area or service consultant's station, many customers appreciated the opportunity to watch the technicians working on their automobiles. They could watch their automobile being brought into the service bay, see how the work was proceeding, and know when the repair was completed. They were assured that the repairs and maintenance work being conducted were, in fact, taking place.

Customers under Contract

Some service facilities have contracts to service the automobiles of other businesses. These contracts vary in terms of the different services the facility provides to the business. In any case, the type of services directly impacts the other duties assigned to the service consultant.

One of the services that an automobile facility may agree to provide to another business may be to pick up, service, and deliver the company vehicles. As a result, the service consultant may have to call the business to schedule the maintenance appointments. In some cases, the maintenance services may be provided in the evening or on a weekend. This requires the service consultant to make arrangements for drivers and technicians to work late or on the weekend.

Contracted services can also require service consultants to make sublet sales arrangements with another business. For example, if a service agreement includes emergency repairs, the service consultant may have to call a towing company to pick up a vehicle. In other words, service contracts with other businesses can add another dimension to a service consultant's position. Service consultants must know what services are to be provided and then have a plan to provide them. Most importantly, service facilities and consultants must not promise something they cannot deliver.

Summary

The "other duties" assigned to a service consultant vary from one type of service facility to another. This cannot be stressed enough because these duties can make a huge difference in a service consultant's job. Some service consultants are the "go to" person. They are at the hub, and daily operations revolve around them. Others are limited to taking care of customers, such as at a large dealership with four or five service consultants. In these cases, other employees take care of many of the other duties while the service consultant works with a lead technician and interacts with the same three or four technicians.

The point in recognizing these other duties is that some people like being the go to person who is at the hub, and others do not. When applying for a service consultant's position, this difference must be carefully considered. A person may not like a service consultant's job, not because of the primary responsibilities of the position but because of the other duties. A person who is not satisfied with his or her job will not be successful. In addition, people cannot take a position and then try to make it fit the way they want to do the job. Employees who try to manipulate a job into what they want will fail.

Rather, because service-consulting positions vary among large and small independent repair facilities, franchises, specialty shops, and dealerships, there is a variety of service-consulting positions. Recognizing these differences and then pursuing a position at a facility most appropriate to one's personal preferences is the correct approach and an opportunity not often given to many people. When there is a match between the type of job desired and the job obtained, the probability of success increases considerably. In addition, the likelihood of creating a positive work environment, which is discussed in Chapter 14, is possible.

Review Questions

Multiple Choice

1. In most general repair shops, which of these is LEAST likely to be a sublet repair?
   A. Windshield replacement
   B. Spark plug replacement
   C. Automatic transmission overhaul
   D. Drive shaft balancing

2. A vehicle is being serviced for the first time at the facility and has had numerous repairs at both dealer and independent shops. Service consultant A suggests telling the customer that the shop's technicians are much more competent than those at the other shops the customer has used in the past. Service consultant B asks the customer to bring in the repair records to help the shop get an idea of the vehicle's history. Who is right?
   A. A only
   B. B only
   C. Both A and B
   D. Neither A nor B

3. Service consultant A says that adding a description of the work performed adds value to the repair. Service consultant B says that the VIN may be used to find part applications, especially when ordering from a dealership parts department. Who is correct?
   A. A only
   B. B only
   C. Both A and B
   D. Neither A nor B

4. Service consultant A says that, like for automobiles, maintenance schedules for shop equipment are printed in the owner's manual. Service consultant B says that, like for automobiles, maintenance schedules should be selected based on the use of the equipment. Who is correct?
   A. A only
   B. B only
   C. Both A and B
   D. Neither A nor B

5. Which of the following is true?
   A. Service consultants must know the return policy of the vendor and how old parts removed during service (called cores) are to be returned to the vendor.
   B. Service consultants must know how to obtain parts from new automobile dealerships. This is because the auto parts stores often do not have all of the parts required to fix newer automobiles.
   C. Petty cash is kept on hand to pay for items that must be purchased from a vendor that does not have a credit arrangement with the service facility.
   D. All of the above

Short Answer Questions

1. What are the different methods to arrange for a sublet sale?
2. Describe the various other duties a service consultant may have to perform.
3. What is petty cash used to purchase?
4. Under what circumstances will a service facility use liability insurance and workman's compensation?
5. What are the purposes of security systems at a service facility?

CHAPTER 14

A POSITIVE WORK ENVIRONMENT

OBJECTIVES

Upon reading this chapter, you should be able to:

- *Explain how a positive work environment is related to productivity.*

- *Outline steps to set completion expectations (A.2.4).*

- *Define effectiveness and efficiency.*

- *Explain how* effectiveness *and* efficiency *are related to productivity.*

- *Explain why customer expectations, effectiveness, and efficiency are critical to a business (A.2.8).*

- *Describe the basic requirements needed to establish and maintain a positive work environment.*

Introduction

A positive work environment depends on the effectiveness and efficiency of an automotive service facility. The work environment is one of the reasons why people keep working for an employer or leave. The environment can also have an impact on the physical and mental health of employees. In other words, a negative work environment is not productive, is a major reason for high turnover, and can be stressful to the point of being harmful to health of the employees and owners.

One of the responsibilities of a service consultant is to create and maintain a positive environment with the objective of generating quality work and high productivity. The purpose of this chapter is to discuss the connection among work environment, performance, and productivity and how the service consultant can influence them.

Productivity, Effectiveness, and Efficiency

When a skilled and experienced technician works on an automobile and all goes well, he or she may do the work faster than estimated by the labor guide. This causes productivity to go up! When a job does not go well and problems arise, or when a less skilled and experienced technician works on a job, it may take longer than the labor guide allows. When this happens productivity declines! When a job is done in less time than estimated, the profit for a business increases. A job that takes longer than estimated reduces profits, possibly to a point where a loss occurs.

Whether a service facility is productive depends on the effective and efficient performances of the service consultant, parts specialist, and technicians.

- **Effectiveness** requires the estimates, ordering of parts, communication among team members, and the work to be done correctly. Errors reduce productivity, cause customers to be unhappy, and cost the business money!
- **Efficiency** regards whether or not resources of the facility (including labor) are used properly and to their best advantage. For example, too much time spent on processing repair orders, ordering and receiving parts, and completing a job reduces productivity, which reduces the income of the business.

If a job is done effectively (correctly) but is taking too long because it is not done efficiently, it will cost the business more money to do the job than it should. If a job is done quickly and efficiently but not effectively because it has to be done again, the business loses money. Therefore, jobs have to be done both effectively *and* efficiently. For a service facility to make a profit, more jobs must go "right" (effectively and efficiently) than go wrong.

Influences of Effectiveness and Efficiency on the Work Environment

Consequently, when a facility is not effective and efficient because jobs do not go right, productivity declines. When productivity declines, profits turn into losses. The work environment will likely become negative or even hostile. Eventually, the environment and losses threaten the future existence of the facility.

Therefore, the first requirement for a positive environment is for a facility to become more effective and efficient. If productivity and profits improve, worker security improves and morale typically improves as well. To promote effectiveness and efficiency, service consultants must monitor the activities and operations that can influence them. Throughout this book, different methods to monitor various activities and operations and to measure productivity have been discussed, for example, the use of tracking sheets, keeping an eye on flat-rate earnings, collecting customer surveys, keeping a record of comebacks, reducing comebacks, active delivery, and so on.

Improving Effectiveness

When a service consultant has a concern about the quality of work produced by the employees, the solution is for everyone to do their job correctly. Of course, in the case of educated and skilled employees, their effectiveness depends on their knowledge and ability to do their job, especially the technicians. This, in turn, means the facility must hire good people. Because the recruitment and selection of quality personnel is another topic for another book, the assumption here is that the facility has hired employees who can do their job correctly. Therefore, the challenge is to verify that all work is being done effectively.

The most obvious evidence of an ineffective performance is when a customer's automobile has not been properly serviced or has been damaged at the facility and has come back for service. Comebacks represent the most serious problem caused by ineffective performances. There may be a number of reasons for a comeback, and the service consultant must thoroughly review the possibilities. For example, an employee may not have done a job correctly. On the other hand, as discussed in the previous chapters, the comeback may be due to the failure of a part placed on the car. A parts warranty should cover the costs. However, a comeback may be due to the customer refusing to have recommended work done at the time of a repair, such as the case discussed earlier of the woman who did not have the thermostat replaced at the time her cooling system was repaired. In the first case, the comeback would be due to poor workmanship but in the latter two cases, it would not.

Another effectiveness problem occurs when a facility does not identify or gain permission from a customer to perform all of the services needed on an automobile. To have an automobile leave a facility with a problem that was not found represents an ineffective performance on the part of the technician. To have a customer leave a facility with an automobile with a problem that was found, reported, and not called to the customer's attention represents an ineffective performance on the part of the service consultant.

From the customers' standpoint, the examination of their automobile is important because the inspection adds value to their visit. Actually, many customers seek out and have their maintenance work done at service facilities where a technician inspects their automobile, as opposed to a facility where the maintenance work is conducted by a worker who does not have a technical background. Because this inspection does not cost the customer anything, it is referred to as a **value-added service**. Therefore, when customers have their maintenance work done at a service facility, a technician, using a check sheet presented in Chapter 4, should conduct the value-added inspection. Customers, of course, hope that nothing wrong will be found with their automobile but, if there is a problem, they typically welcome the interest shown in keeping their automobile in proper working condition.

Value-added inspections require an effective technician to examine each automobile carefully. Technicians who do not have the ability to examine automobiles carefully and assist the service consultant in generating additional sales must be trained to do so. Technicians who only want to perform the work listed on the repair order and do not examine the automobile should be used as parts installers.

When service consultants recognize the technicians they can rely on to properly conduct the inspections to identify potential problems that need repair, they should select them to form a team. The technicians and service consultant should set procedures so that the service consultant is immediately informed of any problems discovered along with the recommendation of the technicians. The service consultant must then inform the customer as soon as possible about the problem, refer to the repair categories presented in Chapter 9 to describe the severity of the problem, and then use the feature-benefit method of selling, also presented in Chapter 9. If the customer approves the repair, the parts must be ordered immediately so the repair can be made.

Improving Efficiency

Efficiency refers to the economical use of resources, especially human resources. The use of employee time is a primary concern for service consultants, especially for the technicians. The improper use of technicians is economically damaging and, therefore, is evidence of inefficiency. For example, assume the service consultant must ask someone to pick up a part at a local dealership. Assume the service consultant could send either a highly skilled technician or a lower skilled and lower paid maintenance technician. Although the skilled technician can check out the part before bringing it back to the facility, the maintenance technician should always be sent. If a skilled technician is sent and a customer comes in for a diagnosis and repair, a sale could be lost. If the sale is not lost, the customer will have to wait until the skilled technician returns. This is an inefficient use of a resource, which is the technician's time. A general rule is that skilled technicians should never be given work maintenance technicians can perform. Of course, if a facility has only maintenance work, the skilled technicians should be used to get the work out the door.

Efficiency can be monitored more easily than effectiveness. First of all, the average labor sales and average parts sales per automobile for a set period of time (such as a week or a month) should be calculated. This average is determined by dividing the gross labor (or parts) sales for the given period of time by the number of automobiles serviced. For example, if the gross labor sales for the week were $1200 and twenty automobiles were serviced, the average labor sale per vehicle was $60 ($1200/20 vehicles = $60 per vehicle). This calculation can also be made for each technician. In addition, a facility may separate its maintenance sales from its repair sales and then make the average sale calculations.

The same calculations can be made for the sale of flat-rate hours to avoid using dollars as a measure. In these cases, the gross labor sale per automobile would be converted to labor (flat rate) hours by dividing the average gross sales per automobile by the labor rate. For example, if the labor rate is $37.50 per hour, the number of flat-rate hours sold per automobile is 1.6 hours ($60/$37.50 = 1.6 flat-rate hours sold per automobile). The labor rate sales per technician can also be calculated.

With respect to parts, assume that for the twenty automobiles, the gross parts sales were $600. This means that an average of $30 in parts was sold per automobile. Adding the average labor and parts sales, the weekly total sales per automobile would be $90 ($60 in labor plus $30 in parts).

The question is whether the total sales can meet expenses. In these reviews, the hourly rates of the technicians can also be compared to the labor sales. If the sales do not cover expenses, the facility must increase sales and production, be more efficient, or both. For example, if an increase in sales is needed, the number needed is used as a target for improvement. A danger, however, is that a technician or team may be tempted to oversell.

Employee Input for Improvement

Employee feedback and suggestions for the improvement of efficiency and effectiveness should be encouraged since each employee's financial future rides on keeping his or her job. Each person, therefore, has a vested interest in making the business more effective and efficient. Some companies believe that employee feedback is so important that they formally survey employees on a regular basis. In some of these cases, groups of employees are brought together to identify ways the business can be improved.

The purpose of employee surveys and group meetings is to identify current, emerging, and potential future problems that can affect the facility's immediate as well as its long-term survival. For example, when a problem is identified, employee participation to help solve the problem can be beneficial. In addition, the employees often become more sensitive to any activities that can hinder the effective and efficient operations of the business. Not only might they advise management of the problems they see but they can often come up with solutions to fix them.

Customer Expectations

A positive work environment is also dependent on meeting the expectations of customers. As stressed in the previous chapters, when customers are satisfied, they will return for additional service and will recommend the facility to their friends. Their satisfaction is important because facilities depend on repeat customers, as well as a steady stream of new customers, to maintain profitability. One of the best sources of new customers is the referrals of current customers.

Identifying Completion Expectations of Customers

Unfortunately, the personal expectations of customers are hard to generalize about because of their different needs and backgrounds. As a result, service consultants must attempt to predict customer expectations while they are determining what has to be done to the automobile. This, obviously, is difficult since many customers do not know how automobiles work or what expectations are reasonable and are skeptical about service facility recommendations. Furthermore, the expectations of customers who see their automobiles as simply a means of transportation will likely differ from those who have a personal attachment to them.

Service consultants, therefore, must interview the customers to gain an idea of their knowledge and perceptions and to clarify their expectations. Then they must assure them that their expectations will be met or inform them that they are not realistic. At the same time, service consultants must instill confidence in customers that the service recommended will solve their problems and that they will be properly and promptly performed.

After working with the customer, the service consultant gives the work order to the technician as described in the earlier chapters. At the same time, however, the service consultant should communicate the information gathered about the customer to the service team. In some cases, this additional bit of information can be important to the team's attainment of its common goal, which is to meet customer expectations. For example, one customer at Renrag Auto Repair had his oil changed regularly. Because of his extensive travels, he had his wife call on the afternoon he would come in. He always arrived at 4:30 and wanted a specific type and brand of oil, which was not kept on inventory. As a result, a case of this oil was stored in the supply room specifically for this customer. When he came in, extra effort was always put forth to have his car serviced at the time of his arrival. While the oil sales alone might not justify the extra effort, the repairs on his automobile and his referrals did.

Encouraging Efforts to Exceed Expectations

Long-term success cannot be achieved by a slick program, gimmick, or expensive improvements. It ultimately comes down to the effectiveness and efficiency of the technicians, service consultant, manager, and owner as well as their attitude about meeting customer expectations. This positive attitude can translate into a positive environment, which is often what really counts when it comes to customer satisfaction. No one likes going into a business where the personnel are negative or unfriendly, no matter how well they do their job.

Positive attitudes, a positive work environment, effectiveness, and efficiency cannot be bought. In other words, money will not buy better employee attitudes or even better performances. Rather, when employees prove they can do the job and that they care about meeting or exceeding customer expectations, reward bonuses for improved performance can be considered. A company policy to grant bonuses, however, is a complex topic that is a managerial prerogative.

Therefore, because bonuses are a company policy, service consultants must use other methods to encourage employees to meet goals and conduct work that meets or exceeds customer expectations. When expectations are met or exceeded, they must be recognized via a compliment to the employee or a letter to the manager and/or owner. At the same time, when expectations are not met, they too must be informally and/or formally recognized. This relatively simple practice can have an effect on the work environment. Too often employees do not care because they do not believe that other people do.

The Follow-Through

After a repair or maintenance has been completed, service consultants must follow through and ensure all services were provided, recommendations for additional services were reported to the customers, and the correct procedures for preparing the automobile for delivery were conducted, such as ensuring the vehicle is clean and free of shop materials. To assist the service consultant and technician in preparing the automobile for this active delivery, a "checkoff" sheet should be attached to every repair order (see Chapter 9). These sheets, when properly designed and used, should guarantee that customer expectations have been met.

At the time of delivery, the service consultant must also translate any of the technician's comments on the repair order into words the customer can understand. As suggested above, this should be done to support the environment by instilling customer confidence that the recommended service solved his or her problem and was properly performed. For example, a technician who worked on an automobile that did not idle correctly wrote:

> "Followed service manual steps and checked the base idle and timing. Adjusted base timing to 10 degrees BTDC. Desludged throttle body and set base idle to 600 rpm. Pulled computer codes and all passed. Checked TP voltage and found it was at 2 volts. Tried to adjust TP sensor but still out of range. Replaced TPS and was able to reset TP voltage to within 1/10 volt of manufacturer's 1 volt specification. Performed idle learning strategy procedure, road tested car, and verified idle was normal."

This detailed comment is possibly more than some technicians would write; however, it illustrates how a very detailed comment cannot be presented to a customer on an invoice. It is too long and far too technical for the customer to understand. Rather, this comment could be explained to the customer as follows:

> "Technician verified concern. Parts related to the computer controlled idle were inspected, adjusted, and tested. A defective throttle position sensor was replaced and idle was verified to operate properly."

This translation of the technician's comment will help the customer to understand what was done to fix the problem. Another possibility is to show the customer the old parts and to offer to give them to the customer if there is no core charge or if it is not leaking fluids. At Renrag Auto Repair, when old parts were offered, most customers did not take them home.

If a customer chooses to have more specific information on a repair, then the technician's comments should be shown to him or her. In some cases the technician may be asked to explain the repair to the customer. At Renrag Auto Repair, customers were quite impressed when the technician who actually performed the repair explained the cause and cure in person. This personal touch also showed customers that a team of employees was looking after their best interests.

What to Say to Avoid Problems

Meeting customer expectations does not mean that service consultants should comply with every request. There are times when a customer request must be denied; however, the denial must be justified. To arrive at a "yes" or "no" answer, there are five questions service consultants should ask themselves. These questions are demonstrated in the following example.

Fred brings in a new chemical additive that, according to the chemical manufacturer, will do great things for his transmission. He requests that the technician pour this mystery liquid into his automatic transmission when the technician services it. Before agreeing to the customer's request, the service consultant should ask these five basic questions.

Will pouring the mystery liquid into the transmission:

Help the customer? Yes. It is nice to be agreeable. Pouring in the liquid might help the customer because he will not have to pour it in himself.

Promote sales? No. The business did not sell him the liquid and the act of pouring it in does not mean the business will generate any additional sales. After all, do people go to a restaurant with their own raw hamburger and ask the restaurant to cook it? Probably not, and even if they did, the restaurant probably would not cook it because it has no idea whether the meat is good quality and was handled properly. Therefore, it just makes good sense for the restaurant to use its own meat for the hamburger. The restaurant managers will know that the beef will taste good and will not make customers sick. In the same way, the service facility spends a great deal of time to secure high quality parts and products for its customers at a reasonable price.

Protect the company from risk? No. It is unknown what this liquid is and whether it can be used in the transmission. The service facility could be at risk for pouring in the mystery liquid without a written waiver from the customer stating that if it damages the transmission, the business is not responsible. Of course, even with the waiver, the court may view the business as the "automotive experts." In other words, if the liquid caused any damage, the service facility experts should have known better. Thus, the business may still be liable.

Ensure health and safety? No. If there is an accident and the technician spills the liquid on his skin or gets it in his or her eyes, no one will know how to help. Also, if the technician accidentally spills the liquid on the automobile's paint, damage could result. The liquid could even be explosive if handled incorrectly or could damage the transmission's internal parts once inside. Therefore, whether the liquid is safe for the company's technician to handle or use inside the transmission is unknown.

Make a fair and honest profit? No. The business will not profit from pouring the liquid into the transmission.

A "no" response to any one of these five questions indicates that the request should be denied. Since there are so many "no" responses in the example, it would be best to use the one that is the most logical to defend. In this case, declining the request based on "promoting sales" or "making a fair and honest profit" is not relevant to the interests of the customer. In fact, the customer may be offended that his or her "favor" is rejected. "Ensuring health and safety" is a stronger response, but using the "protect the company from risk" answer is the likeliest to defend. The service consultant could deny the request because it is against company policy to pour products into the transmission that are not recommended by the automobile manufacturer.

Summary

Productivity and a positive work environment are essential to the survival of an automotive service facility. As the chapter points out, productivity is dependent on effective and efficient service and customer satisfaction. At the same time, a positive work environment is needed to keep effective and efficient employees who are willing to take on hard and demanding automotive service work.

When a facility is operated properly, it will be profitable and employees will be well paid. When it is not operated properly, the organizational structure (see Chapter 1) breaks down, the work environment is not positive, profitability is limited or nonexistent, and the life of the business is limited.

One of the key people, if not the most important person, who makes a service facility productive is the service consultant. Too often this person is thought of as a clerk or salesperson. As the text demonstrates, this is not the case nor should be the case. There are many dimensions to the service consultant's job. They must not only perform all of the tasks covered in this book, but they must also concern themselves with meeting sales targets, keeping customers satisfied, and motivating their team.

In addition, the service consultant must be a leader. Although the organizational diagrams may not place this person in a position with leadership authority, it typically comes with the job. A service consultant must get pleasure from being busy, enjoy working with the public, like being around automobiles, and be willing to accept leadership responsibilities.

Review Questions

Multiple Choice

1. A technician turns in a repair order that recommends replacement of the CV boot with no further description. Which of these should the service consultant do next?
   A. Estimate replacement of the complete axle.
   B. Verify parts availability.
   C. Determine the reason for the repair.
   D. Check the vehicle repair history.

2. When a customer is picking up his or her vehicle, Service consultant A says that it is important to take the time to explain the work performed in as much detail as the customer requires. Service consultant B says that if the customer asks questions it indicates that he or she does not trust the shop/dealership. Who is correct?
   A. A only
   B. B only
   C. Both A and B
   D. Neither A nor B

3. A customer calls for a service consultant who is already working with a customer. Which of these should the service consultant do?
   A. Take the customer's name and number and promise to call back.
   B. Attempt to help the customer.
   C. Place the customer on hold until the service consultant becomes available.
   D. Transfer the call to the owner/service manager.

4. A service consultant has prepared an estimate from a technician's diagnosis. Before providing the customer with the estimate, which of these should the service consultant perform first?
   A. Agree on a completion time with the technician.
   B. Verify availability of necessary parts.
   C. Perform a thorough test drive.
   D. Identify additional maintenance needs.

5. A fellow service consultant is upset with one of the shop's technicians. Which of these should the service consultant do?
   A. Encourage the other service consultant to talk with the technician.
   B. Offer to work with that technician until the situation blows over.
   C. Speak to the technician.
   D. Alert the service manager immediately.

6. A customer has come to pick up his or her vehicle when the service department is very busy. Which of these is the best way to handle the situation?
   A. Direct the customer to the cashier to cash out.
   B. Advise the customer that you are very busy.
   C. Review the work performed and the invoice with the customer.
   D. Ask the customer to come back when it is quieter.

Short Answer Questions

1. How can a service consultant create a positive work environment, and how will this help productivity?
2. What are completion expectations?
3. Define effectiveness and efficiency.
4. How are effectiveness and efficiency related to productivity?
5. How are customer expectations related to effectiveness and efficiency?
6. What are the basic requirements a service facility needs to establish and maintain a positive work environment?

PART IV

CLINICAL PRACTICUM EXERCISE

After reading the information in Chapters 13 and 14, visit a service facility and interview a service consultant or manager. Ask what "documents" he or she uses to run the business. Specifically think in terms of the following categories:

- PRODUCTIVITY DOCUMENTS: Technician time tickets and individual and team targets for improvement such as "average labor sale per vehicle."
- WORKFLOW DOCUMENTS: Repair orders and service department customer tracking sheets.
- FINAL WORK INSPECTION: Process inspection reports such as technician checkoff sheets, technician comments on the repair order, government agency records, or CSI ratings.
- SUPPLIER DOCUMENTATION: Parts supplier delivery receipts and subcontractor invoices.
- FEEDBACK DOCUMENTATION: Customer and employee feedback and how it changes the facility's customer service procedures. NOTE: When possible, ask to see examples of customer satisfaction surveys and ask how they are distributed.
- OTHER DOCUMENTATION: Customer repair order files and facility security checkoff sheets.

Small Group Breakout Exercises

Activity 1: Discuss customer satisfaction in small groups. Plan customer satisfaction improvements that will cost little, if any, money for a facility to implement. Share your ideas with the class.

Activity 2: Examine some of your school's shop equipment and, using the owner's manual when available, determine the maintenance requirements of the equipment. In a small group make a list of all of the equipment examined and what type is is as well as when maintenance must be completed.

Activity 3: Make a list of all of the special tools in your school's lab. Examine or create a method to mark, box, tag, or hang the tools (when possible, cross-reference the special tool number to manufacturer shop manual information) so it can be found more efficiently. Then evaluate the method to determine whether it would be practical to use in a service facility. If so, what procedures must be implemented to ensure the system will work?

APPENDIX

ASE C1 Task List: Service Consultant

A. COMMUNICATIONS
A.1. CUSTOMER RELATIONS
A.1.1 Demonstrate proper telephone skills.
A.1.2 Obtain and document pertinent vehicle information and confirm accuracy.
A.1.3 Identify customer concern/request.
A.1.4 Obtain and document customer contact information.
A.1.5 Open repair order and confirm accuracy.
A.1.6 Demonstrate appropriate greeting skills.
A.1.7 Arrange for alternate transportation.
A.1.8 Promote procedures, benefits, and capabilities of service facility.
A.1.9 Check vehicle service history.
A.1.10 Identify and recommend service and maintenance needs.
A.1.11 Communicate completion expectations.
A.1.12 Obtain repair authorization.
A.1.13 Identify customer types (first time, warranty, repeat repair, fleet, etc.).
A.1.14 Present professional image.
A.1.15 Perform customer follow-up.
A.1.16 Explain and confirm understanding of work performed, and charges; review methods of payment.

A.2. INTERNAL RELATIONS
A.2.1 Effectively communicate customer service concern/request.
A.2.2 Understand the technician's diagnosis and service recommendation.
A.2.3 Verify availability of required repair parts.
A.2.4 Establish completion expectations.
A.2.5 Monitor repair progress.
A.2.6 Interpret and clarify repair procedures.
A.2.7 Document information about services performed or recommended.
A.2.8 Maintain shop production and efficiency.
A.2.9 Maintain open lines of communication within the organization.

B. PRODUCT KNOWLEDGE
SECTIONS B.1 THROUGH B.4 HAVE THE SAME THREE TASKS:
TASK 1) Identify major components and location
TASK 2) Identify component function
TASK 3) Identify related systems

B.1.   ENGINE SYSTEMS — Tasks 1, 2, and 3 (includes mechanical, cooling, fuel, ignition, exhaust, emissions control, and starting/charging).

B.2.   DRIVETRAIN SYSTEMS — Tasks 1, 2, and 3 (includes manual transmission/transaxles, automatic transmission/transaxles, and drivetrain components).

B.3.   CHASSIS SYSTEMS — Tasks 1, 2, and 3 (includes brakes, suspension, and steering).

B.4.   BODY SYSTEMS — Tasks 1, 2, and 3 (includes heating and air conditioning, electrical, restraint, and accessories).

B.5.   SERVICES/MAINTENANCE INTERVALS
    B.5.1   Understand the elements of a maintenance procedure.
    B.5.2   Identify related maintenance procedure items.
    B.5.3   Locate and interpret maintenance schedule information.

B.6.   WARRANTY, SERVICE CONTRACTS, SERVICE BULLETINS, and CAMPAIGNS/RECALLS
    B.6.1   Demonstrate knowledge of warranty policies and procedures/parameters.
    B.6.2   Locate and use reference information for warranties, service contracts, and campaigns/recalls.
    B.6.3   Explain warranty, service contract, technical service bulletin, and campaign/recall procedures to customers.
    B.6.4   Verify the applicability of warranty, service contract, technical service bulletin, and campaigns/recalls.

B.7.   VEHICLE IDENTIFICATION
    B.7.1   Locate and utilize vehicle identification number (VIN).
    B.7.2   Locate the production date.
    B.7.3   Locate and utilize component identification data.
    B.7.4   Identify body styles.
    B.7.5   Locate paint and trim codes.

C. SALES SKILLS
C.1.   Provide and explain estimates.
C.2.   Identify and prioritize vehicle needs.
C.3.   Address customer concerns.
C.4.   Communicate the value of performing related and additional services.
C.5.   Explain product/service features and benefits.
C.6.   Overcome objections.
C.7.   Close the sale.

D. SHOP OPERATIONS
D.1.   Manage workflow.
D.2.   Use available shop management systems (computerized and manual).
D.3.   Identify labor operations.
D.4.   Demonstrate knowledge of sublet procedures.
D.5.   Maintain customer appointment log.
D.6.   Address repeat repairs/comebacks.

GLOSSARY

Active delivery—when a service consultant personally delivers an automobile to a customer to discuss matters of importance, to check the condition of the automobile, and to determine customer satisfaction.

Automobile service facility—a for-profit business that performs the maintenance, repair, and diagnosis of automobiles.

Bumper-to-bumper warranty—a contract that covers all of the components in an automobile (see *warranty contract*).

Campaign repairs—see *recall/campaign repairs*.

Chain service facility—an automobile service facility that is one of several facilities owned by a corporation, such as Pep Boys and Sears, Roebuck, & Co.

Closing a sale—when a customer gives an approval for a service by signing a repair order.

Comebacks—occur when a customer had a repair made to an automobile and must return to have the same repair made again.

Company policies—statements that indicate how the owners want to conduct business and legal regulations that direct the way business and services must be conducted.

Cores—a used part that has been replaced and must be returned to the supplier for a reduction in the cost of the new part.

Corporation—a business owned by one or more people who invest in the business by purchasing shares of stock.

Cross training—when workers are taught how to perform one or more jobs in a business.

Customer automobile inventory sheet—a form used to record customer automobiles located in the building and on the property of the facility.

Customer status sheet—a form used to record customers' automobiles being serviced and the status of these services, from the creation of the repair order to the payment of the invoice.

Daily customer log—a calendar with pages to enter daily appointments, vital information on each customer, services to be performed, and the estimated time needed for each service.

Effectiveness—when work is done correctly and quality outcomes are achieved.

Efficiency—when resources are used properly and output is maximized.

Emission warranty—a federal requirement that new automobile manufacturers must guarantee the repair of their automobile emission components for a stated period of time.

Extended warranty—purchased by owners of new and used automobiles to provide warranty coverage (sometimes with a deductible) for identified components in an automobile for a given period of time or number of miles that the automobile is driven from the date of purchase.

Feature-benefit selling—a sales approach that explains the service to be provided (feature) and its advantage (benefit).

Flat-rate objective—the number of flat-rate hours a technician hopes to earn in a week.

Flat-rate or Flat-rate pay system—Technicians who are allotted (and paid for)and paid for a specified amount of time to perform a job are considered flat-rate technicians and are paid on the flat-rate pay system.

Fleet service facility—a service facility that limits its services to vehicles owned by a company or government body.

Flow diagram—a drawing that illustrates the processing of work or an activity through a business or an organization.

Formatted system—when an operations manual is followed to encourage consistency in performances of an automotive repair facility.

Franchise—a business granted the use of a nationally recognized name in return for a fee and a percentage of sales.

Garage keepers insurance—liability insurance that covers damages to customers' automobiles while at a facility.

Gross profit—the balance left over from gross sales after subtracting technician salaries and the cost of the parts used to repair customer automobiles.

Gross sales—the total amount of money received from customer sales.

Invoice—final bill (see vendor invoice).

Job description—a list or description of the job tasks to be performed by a person in a position.

Job duties—details of job tasks.

Job tasks—major work assignments conducted by a person in a specific job.

Lead technician—a technician who assists with the coordination of the work in a shop, in assigning jobs to other technicians, in quality control, in monitoring the condition and use of equipment, and with other supervisory duties assigned by management.

Lemon law—a law that requires automobile manufacturers, and, in some states, the automobile dealers, to buy back an automobile when it is not properly repaired.

Maintenance contract—awarded to or purchased by a customer at the time an automobile is purchased and pays for specified maintenance services for a set period of time from the date of purchase or for a set number of miles an automobile is driven.

Markup—the difference between the amount a facility pays for a part and the amount it is sold to a customer.

Mechanic's lien—state laws that permit a service facility to hold a customer's automobile until payment for a service is received.

Net loss—a negative balance after expenses and business taxes are deducted from the gross profit.

Net profit—a positive balance after expenses and business taxes are deducted from the gross profit.

New automobile manufacturer warranty—a contract awarded to the owners of an automobile that provide for its repair at no charge for a predetermined length of time since its initial purchase or for a predetermined number of miles the automobile is driven.

Open-business environment—forces outside the control of a service facility that affect its sales potential.

Operational environment—features or forces that have an influence on the daily business activities of a service facility.

Operational procedures and regulations—rules based on company policy.

Operations manual—directs the way work is to be conducted and processed in each work area.

Organizational structure—represents the managerial chain of command and the relationships among its positions.

Overselling—when customers are sold a service that their automobile does not need.

Partnership—a business owned by two or more people.

Petty cash fund—a fund that makes cash available for the purchase of items below a specified amount of money.

Policy check—a check given to customers for the purchase of goods or services from the service facility.

Pre-priced maintenance menu—a chart presenting different maintenances suggested by a manufacturer and the charges for each.

Product-specific service facility—a service facility that diagnoses, repairs, and performs maintenance on specific makes and models of automobiles, such as a facility that services only Volkswagens.

Proprietorship—a business owned by one person.

Recall/campaign repairs—when a manufacturer requests the owners of a specific year, make, and model of an automobile to take it to a service facility for repair at no charge to them.

Repair categories—a list used to indicate the relative importance of a suggested repair.

Repair order tracking sheet—a form used to monitor the status of each repair order.

Seamless system—a process that is not disrupted when work is passed from one person or one station to another.

Service—the maintenance, repair, and diagnosis of an automobile.

Shop management system—the procedures, documents, files, and computer used to prepare, store, and retrieve customer information, repair orders, automobiles serviced, flat rate hours, parts markups, parts inventory, and vendor information.

Specialty service facility—services specific makes and models of new and/or used automobiles, such as those sold by a dealership.

Specialty team—a team whose repairs, maintenance, or diagnostic work assignments are limited, such as a team that performs only maintenance work.

Stockholders—people who invest in a corporation by purchasing shares of stock in the business.

Sublet sales—when a facility sends a customer's automobile to another facility for service, has the automobile returned when completed, pays the other facility for the service, and charges the customer for the service conducted by the other facility.

System-specific service facility—a service facility that repairs and maintains one automobile system, such as transmissions or brakes.

Tactical environment—factors that influence the provision and supply of resources needed to conduct business activities.

Technical service bulletin (TSB)—an announcements put out by automobile manufacturers to dealerships and subscribers of computer repair information systems to announce automobile operational concerns and how to repair them.

Technician's hardcopy—a thicker, cardboard-like copy of a repair order given to the technician.

Up-selling—when a customer selects higher-quality parts and/or labor as opposed to less expensive parts and/or labor.

Value-added service—a service received by a customer at no extra cost.

Vehicle identification number (VIN)—a seventeen-digit number assigned to an automobile when manufactured.

Vendor invoice—a bill from another business for parts, equipment, or services purchased.

Vendors—a company that sells parts, goods, and services to a facility.

Warranty contract—if a part does not work properly or breaks within a set period of time, a new part, and possibly the cost of the labor to replace it, will be provided at no charge to the customer.

Workflow—the processing of work from the initial contact with customers to the return of their automobile.

Workers' compensation—an insurance policy that covers injuries received by employees while at work.

INDEX

A

Active delivery, 137–138
  actual delivery, 172
  care and cleanliness of vehicle, 172
  follow-up call, 138, 179–181
  future appointment, 138
  needed services, 137
  positive points, 137
  pre- and post-inspection of vehicles, 171
  workflow process, 171–172
Advertising, 185–186, 206
  balanced advertising plan, 185
  methods of, 185
Alternative transportation, 118, 131
Answering customer questions, 99
Appointment, customer, 65, 66
  appointment book recorded, 163, 164
  computerized record, 163
  as part of workflow process, 162
  time required for, 164
ASE certification, 17–18
ASE. See National Institute for Automotive
    Service Excellence.
Automobile dealership, 12–13
  chain of command, 13
  definition, 12–13
  flowchart, 13
Automobile service facility, 4
  general requirements, 6
  guidelines for, 14–16
  history of services performed, 60
  objective of, 4
  purpose of, 4
  service consultant requirements, 6
  tracking earnings and losses, 60

Automobile service facility, types of:
  chain, 12
  corporate owned, 9
  dealership, 12–13. See also product-specific.
  fleet service, 14
  franchise, 10
  privately owned, 8
  product-specific, 6–7
Automobile service history, 62
  customer approval, 62
  computer database, 62
  general maintenance inspection form, 63,
    135
  invoices, 62
  repair orders, 62
  special maintenance inspection form, 64,
    135
Automotive service, historical discussion of,
    40–41

B

Balance, 18
Banking, 205–206
Billable hours, 42, 50
  calculation of, 42–43
Bonuses, 217
Bumper-to-bumper warranty, 79
Business contracts, 200–201
  goods, resale of, 202–203
  maintenance contracts
    for building, 202
    for equipment, 202
  parts, supply of, 202–203
  uniforms, 202
  sublet sales, 201